I0465725

PUBLISHER COMMENTARY

We print NASA's handbooks and standards for the convenience of those that use them on a daily basis. We print all of these a full 8 ½ by 11 with large text so they are easy to read. Yes, color books are expensive to print so unless the information relies on the use of color for proper interpretation or understanding, we print most books in black and white to keep the cost down. All these documents are available for download for free from NASA, however printing them all over a network printer would take days.

Why buy a book you can download free? We print this so you don't have to.

All these books are available for free download from the government web site. Some are available only in electronic media. Some online docs are missing pages or barely legible.

We at 4th Watch Publishing are former government employees, so we know how government employees actually use the standards. When a new standard is released, an engineer prints it out, punches holes and puts it in a 3-ring binder. While this is not a big deal for a 5 or 10-page document, many NIST documents are over 100 pages and printing a large document is a time-consuming effort. So, an engineer that's paid $75 an hour is spending hours simply printing out the tools needed to do the job. That's time that could be better spent doing engineering. We publish these documents so engineers can focus on what they were hired to do – engineering. It's much more cost-effective to just order the latest version from Amazon.com

If there is a standard you would like published, let us know. Our web site is www.usgovpub.com

www.usgovpub.com

List of Other NASA Publications Available on Amazon.com:

NASA-STD-5001B	Structural Design and Test Factors of Safety for Spaceflight Hardware
NASA-STD-5006A	General Welding Requirements for Aerospace Materials
NASA-STD-5008B	Protective Coating of Carbon Steel, Stainless Steel, and Aluminum on Launch Structures, Facilities, and Ground Support Equipment
NASA-STD-5009A	Nondestructive Evaluation Requirements for Fracture-Critical Metallic Components
NASA-STD-5012B	Strength and Life Assessment Requirements for Liquid-Fueled Space Propulsion System Engines
NASA-STD-5019A	Fracture Control Requirements for Spaceflight Hardware
NASA-STD-5005D	Standard for The Design and Fabrication of Ground Support Equipment
NASA-HDBK-8739.21	Workmanship Manual for Electrostatic Discharge Control
NASA-HDBK 8739.23A	NASA Complex Electronics Handbook for Assurance Professionals (Color)
NASA-HDBK-8719.14	Handbook for Limiting Orbital Debris (Color)
NASA-HDBK-8709.22	Safety and Mission Assurance Acronyms, Abbreviations, and Definitions
NASA-HDBK-7009	NASA Handbook for Models and Simulations: An Implementation Guide For NASA-STD-7009 (Color)
NASA-HDBK-8739.19-2	Measuring and Test Equipment Specifications NASA Measurement Quality Assurance Handbook – Annex 2
NASA-HDBK-8739.19-3	Measurement Uncertainty Analysis Principles and Methods NASA Measurement Quality Assurance Handbook – Annex 3
NASA-HDBK-8739.19-4	Estimation and Evaluation of Measurement Decision Risk NASA Measurement Quality Assurance Handbook – Annex 4
NASA RCM	Reliability-Centered Maintenance Guide for Facilities and Collateral Equipment

www.usgovpub.com

NOT MEASUREMENT
SENSITIVE

NASA TECHNICAL STANDARD

National Aeronautics and Space Administration

**NASA-STD-5019A
w/CHANGE 2:
ADMINISTRATIVE
CHANGE
2018-03-29**

Approved: 2016-02-01
Superseding NASA-STD-5019

FRACTURE CONTROL REQUIREMENTS
FOR SPACEFLIGHT HARDWARE

DOCUMENT HISTORY LOG

Status	Document Revision	Change Number	Approval Date	Description
Interim	I		2006-09-12	Interim Release
Baseline			2008-01-07	Baseline Release—Transitioned from NASA-STD-(I)-5019
Revision	A		2016-02-01	General Revision. Revised to include only the key and sufficient requirements for fracture control and to incorporate nonmetal requirements that were previously referenced in MSFC-RQMT-3479, Fracture Control Requirements for Composite and Bonded Vehicle and Payload Structures.
		1	2016-02-26	Administrative Change: Corrected section 6.2.5.a(2) from "Metallic parts have a material property ratio of $K_{Ic}/F_{ty} < 1.66$ $\sqrt{mm}$ (0.33 $\sqrt{in}$) and do not have sensitivity to EAC, SLC, or stress corrosion cracking as defined in NASA-STD-6016." To "Metallic parts have a material property ratio of $K_{Ic}/F_{ty} \geq 1.66$ $\sqrt{mm}$ (0.33 $\sqrt{in}$) and do not have sensitivity to EAC, SLC, or stress corrosion cracking as defined in NASA-STD-6016."
		2	2018-03-29	Editorial/Administrative Change: Corrected section 6.1.2.2, NFC External Shatterable Components, to read to: ". . . meet 6.1.2.2.a, 6.1.2.2.b, and 6.1.2.2.c" rather than ". . . meet either 6.1.2.2.a, 6.1.2.2.b, or 6.1.2.2.c."

FOREWORD

This NASA Technical Standard is published by the National Aeronautics and Space Administration (NASA) to provide uniform engineering and technical requirements for processes, procedures, practices, and methods that have been endorsed as standard for NASA programs and projects, including requirements for selection, application, and design criteria of an item.

This NASA Technical Standard is approved for use by NASA Headquarters and NASA Centers and Facilities and may be cited in contract, program, and other Agency documents as a technical requirement. It may also apply to the Jet Propulsion Laboratory and other contractors only to the extent specified or referenced in applicable contracts.

This NASA Technical Standard establishes the fracture control requirements for human-rated spaceflight, since NASA policy states that fracture control be imposed on all human-rated spaceflight hardware. It was developed by a NASA-wide Fracture Control Working Group to provide a common framework for fracture control practices on NASA programs.

Requests for information should be submitted via "Feedback" at https://standards.nasa.gov. Requests for changes to this NASA Technical Standard should be submitted via MSFC Form 4657, Change Request for a NASA Engineering Standard.

__Original Signed By__	__01/07/2008__
Ralph R. Roe, Jr.	Approval Date
NASA Chief Engineer	

TABLE OF CONTENTS

TABLE OF CONTENTS (Continued)

APPENDICES

LIST OF FIGURES

LIST OF TABLES

FRACTURE CONTROL REQUIREMENTS FOR SPACEFLIGHT HARDWARE

1. SCOPE

1.1 Purpose

The purpose of this NASA Technical Standard is to establish the fracture control requirements for National Aeronautics and Space Administration (NASA) human-rated spaceflight hardware. In accordance with NASA Procedural Requirements (NPR) 8705.2B, Human-Rating Requirements for Space Systems, it is NASA's policy to produce human-rated space systems that have failure tolerance for catastrophic events or that potentially catastrophic hazards are controlled through a defined process in which approved standards and margins are implemented that account for the absence of failure tolerance.

This NASA Technical Standard supersedes the baseline release of NASA-STD-5019, Fracture Control Requirements for Spaceflight Hardware.

Programs that are not human-rated may choose to impose these requirements on a mission or hardware to bolster the program or to serve as a stepping-stone for human rating.

1.2 Applicability

This NASA Technical Standard is applicable to human-rated spaceflight hardware.

This NASA Technical Standard is approved for use by NASA Headquarters and NASA Centers and Facilities and may be cited in contract, program, and other Agency documents as a technical requirement. It may also apply to the Jet Propulsion Laboratory and other contractors only to the extent specified or referenced in applicable contracts.

Verifiable requirement statements are numbered and indicated by the word "shall" beginning in section 4. Explanatory or guidance text is indicated in italics beginning in section 4. To facilitate requirements selection and verification by NASA programs and projects, a Requirements Compliance Matrix is provided in Appendix A.

1.3 Tailoring

Tailoring of this NASA Technical Standard for application to a specific program or project shall be formally documented as part of program or project requirements and approved by the responsible Technical Authority in accordance with NPR 7120.5, NASA Space Flight Program and Project Management Requirements.

Technical Authority in this context may vary from program to program. In accordance with NPR 7120.10, Technical Standards for NASA Programs and Projects, section 2.2.4, "The NASA Chief Engineer, the Chief, Safety and Mission Assurance, and the Chief Health and Medical Officer serve as or may delegate Technical Authority for all technical standards within their areas of responsibility."

1.4   Overview

This document provides the hardware developer with the requirements, rationale, and methodologies to implement fracture control requirements. It also provides a guide to the Responsible Fracture Control Board (RFCB) when reviewing the Fracture Control Plan (FCP).

This document contains 26 requirements that are numbered as Fracture Control Requirements [FCRs] and that begin in section 4. These requirements use the word "shall." Narrative text and requirement rationale are provided in italic format beginning in section 4. Narrative text is provided as guidance for the associated requirement. It is recommended that fracture control practitioners become familiar with all portions of this NASA Technical Standard.

The FCRs are summarized and briefly described in table 1, Overview of Fracture Control Requirements in NASA-STD-5019A. Figure 1, NASA-STD-5019A Fracture Control Requirements Diagram, shows a diagram of the FCRs and the section in this NASA Technical Standard in which that particular FCR appears.

A viable fracture control program relies on design, analysis, testing, non-destructive evaluation (NDE), and tracking of fracture critical hardware. It is expected that all spaceflight hardware will be manufactured consistent with industry or aerospace standards, practices, and quality. It is beyond the scope or intent of this document to address technical or quality disciplines that should already exist and be in place regardless of fracture control. Fracture control is imposed and required, not to correct deficiencies in other disciplines, rather to enhance the safety and mission reliability of systems by reducing the risk of catastrophic failure caused by the presence of flaws.

NASA-HDBK-5010, Fracture Control Implementation Handbook for Payloads, Experiments, and Similar Hardware, contains examples and additional guidance for interpretation and implementation of the requirements of this Standard. NASA-HDBK-5010, Revision A, is under development and may not be released at the time of publication of this NASA Technical Standard. Before the release of NASA-HDBK-5010A, the current handbook may provide interim guidance for applying this NASA Technical Standard. Note that NASA-HDBK-5010A will include guidelines for NASA-STD-5019A and will likely undergo a title change to reflect a broader scope than payloads and experiments.

Table 1—Overview of Fracture Control Requirements in NASA-STD-5019A

General Category	Requirement	Description	Section
Overarching fracture control requirement	[FCR 1]	Requires that all hardware developers of human-rated spaceflight hardware implement fracture control by selecting the applicable approaches and activities from sections 5, 6, 7, 8, and 9 of this NASA Technical Standard for all parts and document the applicable FCRs in their hardware-specific FCP for review and approval by the RFCB.	4.1
NASA's implementation of fracture control on human-rated spaceflight hardware	[FCR 2]	Requires implementation by NASA.	4.2.1
	[FCR 3]	Requires implementation by program.	4.2.2
	[FCR 4]	Requires RFCB involvement.	4.2.3
Evaluation of all parts	[FCR 5]	Requires the evaluation of all parts used in human-rated spaceflight hardware for fracture control classification.	4.3
Exempt Classification	[FCR 6]	Criteria for classification of exempt parts	5
Non-Fracture Critical (NFC) Classification	[FCR 7]	Established approaches and activities for NFC parts for specific hardware types	6.1
	[FCR 8]	General approaches and activities for NFC categories	6.2
	[FCR 9]	Additional activities for composite or bonded NFC hardware	6.3
Fracture Critical Classification	[FCR 10]	Criteria for classification of fracture critical parts	7
	[FCR 11]	Established approaches and activities for fracture critical categories for specific hardware types and materials types	7.2
	[FCR 12]	Approaches and activities for metallic hardware types not covered by 7.2 or 7.5	7.3
	[FCR 13]	Approaches and activities for composite or bonded parts not covered by 7.2 or 7.5	7.4
	[FCR 14]	Optional approaches and activities for specific hardware types not covered in 7.2	7.5
Flaw screening and evaluation, traceability, and material requirements for fracture critical parts and other applicable components	[FCR 15]	Flaw screen requirement	8
	[FCR 16]	NDE requirement for metallic parts	8.1.1
	[FCR 17]	NDE requirement for composite or bonded parts	8.1.2
	[FCR 18]	Optional proof test requirement	8.1.3
	[FCR 19]	Optional process control requirement	8.1.4
	[FCR 20]	Detected flaw requirement	8.1.5
	[FCR 21]	Traceability requirement	8.2

General Category	Requirement	Description	Section
	[FCR 22]	Materials requirements	8.3
Documentation	[FCR 23]	Documentation products requirements associated with fracture control	9.1.2
	[FCR 24]	Fracture Control Summary Report documenting all parts	9.1.3
Verification	[FCR 25]	Requirement for verification	9.2
Alternative approaches	[FCR 26]	Requirement for alternative approaches to the requirements of this NASA Technical Standard	10

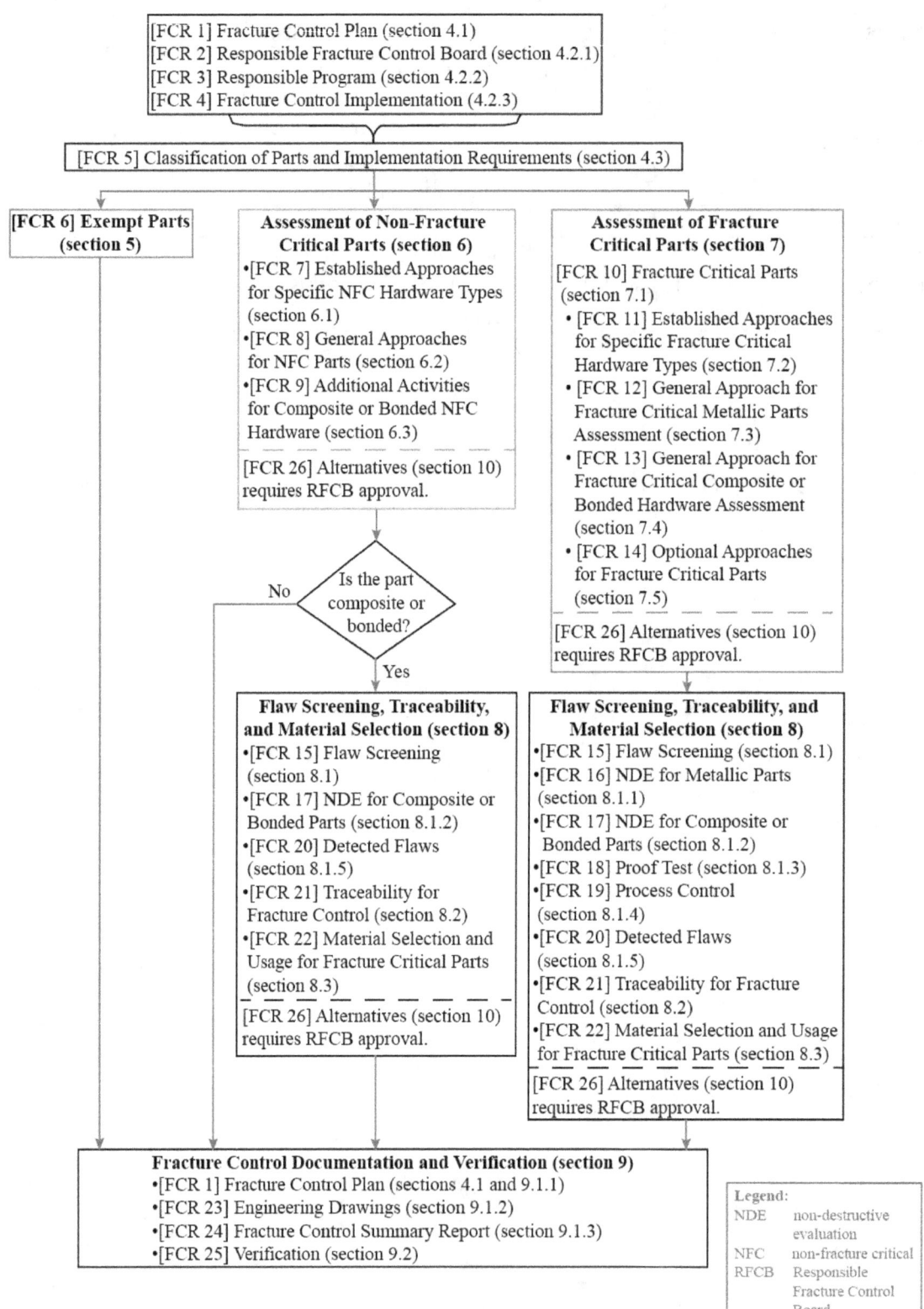

Figure 1—NASA-STD-5019A Fracture Control Requirements Diagram

2. APPLICABLE DOCUMENTS

2.1 General

The documents listed in this section contain provisions that constitute requirements of this NASA Technical Standard as cited in the text.

2.1.1 The latest issuances of cited documents apply unless specific versions are designated.

2.1.2 Non-use of specifically designated versions shall be approved by the responsible Technical Authority.

The applicable documents are accessible at https://standards.nasa.gov, may be obtained directly from the Standards Developing Body or other document distributors, or information for obtaining the document is provided. Reference documents are listed in Appendix B.

2.2 Government Documents

National Aeronautics and Space Administration (NASA)

JSC 20793	Crewed Space Vehicle Battery Safety Requirements
MSFC-STD-3029	Guidelines for the Selection of Metallic Materials for Stress Corrosion Cracking Resistance in Sodium Chloride Environments
NASA-STD-5001	Structural Design and Test Factors of Safety for Spaceflight Hardware
NASA-STD-5009	Nondestructive Evaluation Requirements for Fracture Critical Metallic Components
NASA-STD-5017	Design and Development Requirements for Mechanisms
NASA-STD-5018	Strength Design and Verification Criteria for Glass, Ceramics, and Windows in Human Space Flight Applications
NASA-STD-5020	Requirements for Threaded Fastening Systems in Spaceflight Hardware
NASA-STD-6008	NASA Fastener Procurement, Receiving Inspection, and Storage Practices for Spaceflight Hardware
NASA-STD-6016	Standard Materials and Processes Requirements for Spacecraft

NPR 7120.5	NASA Space Flight Program and Project Management Requirements
NPR 7120.10	Technical Standards for NASA Programs and Projects
NPR 8705.2	Human-Rating Requirements for Space Systems

2.3 Non-Government Documents

Aerospace Industries Association (AIA)/National Aerospace Standards (NAS)

| NASM1312-11 | Fastener Test Methods, Method 11 Tension Fatigue |
| NAM1312-111 | Fastener Test Methods, Metric Method 111 Tension Fatigue |

American National Standards Institute (ANSI)/American Institute of Aeronautics and Astronautics (AIAA)

| ANSI/AIAA S-080-1998 | Space Systems - Metallic Pressure Vessels, Pressurized Structures, and Pressure Components |
| ANSI/AIAA S-081-2000 | Space Systems – Composite Overwrapped Pressure Vessels (COPVs) |

Society of Automotive Engineers (SAE) International

| CMH-17-1G | Composite Materials Handbook |

Southwest Research Institute

NASGRO® User's Manual

2.4 Order of Precedence

2.4.1 The requirements and standard practices established in this NASA Technical Standard do not supersede or waive existing requirements and standard practices found in other Agency documentation.

2.4.2 Conflicts between this NASA Technical Standard and other requirements documents shall be resolved by the responsible Technical Authority.

3.   ACRONYMS AND DEFINITIONS

3.1   Acronyms, Abbreviations, and Symbols

ΔK_{th}	cyclic threshold stress intensity range
ω	maximum operating rotational speed
>	greater than
$\sqrt{}$	square root
®	registered trademark
AIA	Aerospace Industries Association
AIAA	American Institute of Aeronautics and Astronautics
Al	aluminum
ANSI	American National Standards Institute
API	American Petroleum Institute
ASME	The American Society of Mechanical Engineers
ASTM	ASTM International (formerly American Society of Testing and Materials)
atm	atmosphere
BBA	building block approach
BPVC	Boiler and Pressure Vessel Code
cm	centimeter(s)
CMH	Composite Materials Handbook
COPV	composite overwrapped pressure vessel
cp-Ti	commercially pure titanium
CRES	corrosion resistant (steel)
DLL	design limit load
DOT	Department of Transportation
DTA	damage threat assessment
DUL	design ultimate load
EAC	environmentally assisted cracking
ECF	environmental correction factor
EVA	extravehicular activity
F_{su}	ultimate shear strength
F_{tu}	ultimate tensile strength
F_{ty}	yield tensile strength
FAA	Federal Aviation Administration
FCP	Fracture Control Plan
FCR	Fracture Control Requirement
FCSR	Fracture Control Summary Report
ft	foot (feet)
ft- lb	foot-pound(s)

FOD	foreign object debris
HCF	high-cycle fatigue
HDBK	handbook
hr	hour(s)
IDMP	Impact Damage Mitigation Plan
in	inch
J	joule(s)
J_c	critical J-integral
J_{Ic}	plane strain J-integral
JSC	Johnson Space Center
K	stress intensity factor
K_c	plane stress fracture toughness
K_{EAC}	stress intensity factor threshold for EAC in a specific thickness
K_{Ic}	plane strain fracture toughness
K_{Ie}	effective fracture toughness
K_{IEAC}	stress intensity factor threshold for plane strain environmentally assisted cracking
K_{ISCC}	stress intensity factor threshold for plane strain stress corrosion cracking
K_{JIc}	stress intensity factor determined from the plane strain J-integral fracture toughness
K_{SLC}	stress intensity factor threshold for sustained load cracking
kip	kilo-pound
kPa	kilo-pascal
ksi	kip(s) per square inch
LBB	leak-before-burst
LEF	load enhancement factor
LEFM	linear-elastic fracture mechanics
m	meter(s)
mA	milliampere
MDCP	Mechanical Damage Control Plan
MDP	maximum design pressure
MEOP	maximum expected operating pressure
MIL	military
mm	millimeter
MMOD	micro-meteoroid and orbital debris
MPa	megapascal(s)
MSFC	Marshall Space Flight Center
MUA	Materials Usage Agreement
NAS	National Aerospace Standard

NASA	National Aeronautics and Space Administration
NASGRO®	fracture mechanics and fatigue crack growth analysis software
NDE	non-destructive evaluation
NDI	non-destructive inspection
NDT	non-destructive testing
NFC	non-fracture critical
NHLBB	non-hazardous-leak-before-burst
NPR	NASA Procedural Requirements
PRC	process specification
psi	pound(s) per square inch
psia	pound(s) per square inch absolute
RTD	residual threat determination
RFCB	Responsible Fracture Control Board
RQMT	requirement
S	standard
SAE	Society of Automotive Engineers
SLC	sustained load cracking
SPEC	specification
STA	solution treated and aged
STD	standard
Ti	titanium
V	vanadium
ω	maximum operating rotational speed

3.2   Definitions

A-Basis: A statistically calculated number that at least 99 percent of the population of values is expected to equal or exceed with a confidence of 95 percent.[1]

Adhesive Bond (Bond): The joining of parts, components, or materials using a joining substance or agent.

Assembly/Assemblage: An integral arrangement of parts that makes up an individual unit and that acts as a whole.

B-Basis: A statistically calculated value that at least 90 percent of the population is expected to equal or exceed with a confidence of 95 percent.[2]

---

[1] See NASA-STD-6016, Standard Materials and Processes Requirements for Spacecraft; CMH-17, Composite Materials Handbook; Metallic Materials Properties Development and Standardization (MMPDS, Appendix A.2) as appropriate.
[2] See NASA-STD-6016; CMH-17, MMPDS (Appendix A.2) as appropriate.

Bond: The joining of two parts through molecular attraction or through any non-mechanical means of connection.

Bonded Hardware (Structure): Hardware (structure) that is assembled using parts that are joined together with an adhesive.

Brittle Fracture: Sudden rapid fracture under stress (residual or applied) where the material exhibits little or no evidence of ductility or plastic deformation.

Building Block Approach (BBA): A development methodology often used with composites or bonded hardware that (a) starts with selecting the material system and manufacturing approach; (b) moves on to experimentation and analysis of small samples to characterize the system and quantify behavior in the presence of flaws and damage; (c) progresses to examining larger structures to examine buckling behavior, combined loadings, and built-up structures in the presence of credible damage; and (d) finally moves to complicated subcomponents and full-scale components to establish their damage tolerance strength and life. Each step along the way is supported by detailed analysis to validate that the behavior of these structures is well understood and predictable.

Catastrophic Event: Loss of life, disabling injury, or loss of a major national asset.

Catastrophic Failure: A failure that directly results in a catastrophic event.

Catastrophic Hazard: Presence of a risk situation that could directly result in a catastrophic event.

Component: A hardware unit considered a single entity for the purpose of fracture control. A component contains at least one part.

Composite or Bonded Structure: Structure (excluding overwrapped pressure vessels or pressurized components) of fiber/matrix configuration and structure with load-carrying non-metallic bonding agents, such as sandwich structure or bonded structural fittings.

Composite Material: A combination of materials differing in composition or form on a macro scale. The constituents retain their identities in the composite; that is, they do not dissolve or otherwise merge completely into each other, although they act in concert. Normally, the constituents can be physically identified and exhibit an interface between one another. Composite material is not intended to mean an assembly of parts.

Composite Hardware (Structure): Hardware (structure) assembled with parts made from composite materials.

Composite Overwrapped Pressure Vessel: A pressure vessel with a composite structure fully or partially encapsulating a metallic liner. The liner serves as a fluid (gas and/or liquid) permeation barrier and may carry substantial pressure loads. The composite generally carries pressure and environmental loads.

    Contained: A condition in which a suitable housing, container, barrier, restraint, etc., prevents a part or pieces thereof from becoming free bodies if the part or its supports fail.

    Contamination: Any material included within or on the hardware that is not called for on the engineering drawings. Examples of contamination are dust, grease, solvent, solid objects, etc.

    Crack or Crack-like Defect: A discontinuity assumed to behave like a crack for fracture control purposes.

    Critical Stress Intensity Factor: The stress intensity factor at the initiation of crack growth in the part resulting in a catastrophic failure that is representative of the failure mode of concern for the metallic material process condition, weakest orientation, and thickness being evaluated. Examples for metallic materials may include: K_{IEAC}, the stress intensity factor threshold for plane strain environment-assisted cracking; plane strain fracture toughness (K_{Ic}) may be appropriate for thick sections and/or as a lower bound value[3]; effective fracture toughness (K_{Ie}) is used in NASGRO® for crack growth analyses of surface or elliptical flaws; K_{JIc} calculated from J_{Ic} or a K_c calculated from J_c may be appropriate for the conditions described in the defining standard (ASTM E1820, Standard Test Method for Measurement of Fracture Toughness) such as evaluation of ductile tearing and instability; constraint-based assessments (ASTM E2899, Standard Test Method for Measurement of Initiation Toughness on Surface Cracks Under Tension and Bending), and/or tests may be needed for surface or other complex cracks in materials or conditions that invalidate the ability of Linear-Elastic Fracture Mechanics (LEFM) to represent crack growth.

    Damage: See definitions of Flaw and Impact Damage.

    Damage Threat Assessment (DTA): An evaluation of potential sources of flaws in composite or bonded hardware that includes definition, quantification, and an assessment of the residual strength sensitivity to flaws.

    Damage Tolerance: Fracture control design concept under which an undetected flaw or damage (consistent in size with the flaw screening method or residual threat determination (RTD)) is assumed to exist and is shown by fracture mechanics analysis or test not to grow to failure (leak or instability) during the period equal to the service life factor times the service life.

    Design Limit Load (DLL): See definition of Limit Load.

    Design Ultimate Load (DUL): Limit load multiplied by the ultimate factor of safety.

---

[3] Proof test assessments need to use upper bound fracture toughness; see section 8.1.3 in this NASA Technical Standard.

Environmental Correction Factor (ECF): An adjustment factor used to account for differences between the environment (thermal and chemical) in which a part is used and the environment in which it is tested.

Environmentally Assisted Cracking (EAC): A cracking process in which the environment promotes crack growth or higher crack growth rates than would occur without the presence of the environment (ASTM E1681, Standard Test Method for Determining Threshold Stress Intensity Factor for Environment-Assisted Cracking of Metallic Materials). An example is available in published literature (Lewis and Kenny, 1976).

Experiment: For fracture control, an arrangement or assemblage of hardware that is intended to investigate phenomena on a provisional, often human-tended, basis.

Fail-safe: A condition where a redundant load path exists within a part (or hardware), so that after loss of any single individual load path, the remaining load path(s) has sufficient structural capability to withstand the redistributed loads, and the loss of the load path will not cause a catastrophic hazard.

Fastener: For fracture control, any single part that joins other structural elements and transfers loads from one element to another across a joint.

Flaw: For metallics, glass, or brittle materials, a crack-like defect. For composite or bonded materials, an anomaly in the hardware that has the potential for adversely affecting strength, damage tolerance life, or must-work function. Examples of flaws in metallics include cracks, deep scratches and sharp notches that behave like cracks, material inclusions, forging laps, welding incomplete fusion, penetration, and slag or porosity with a crack-like tail. Examples of flaws in composite or bonded materials may include cracks, cuts, scratches, delaminations, porosity/voids, disbonds, wrinkles, foreign object debris, impact damage, etc. Damage (used alone) and flaw are equivalent.

Fleet Leader: Articles representative of spaceflight hardware with respect to production methods, e.g., materials, manufacturing, testing, that either have accumulated (or are scheduled to accumulate) more service lifetime in typical (or more severe) environments than the rest of the fleet and are monitored for indications of failure modes to provide early warning of known and unexpected risks to the rest of the fleet.

Flight (Spaceflight) Hardware: Any hardware (including spares) that is approved to be part of or carried by a launch vehicle, crew module, transfer stage, landing craft, payload, etc.

Flight-like Component: A component assembled and made of parts that are of flight specifications. Flight-like components are usually intended for qualification tests. Any deviations from flight have to be insignificant with respect to test objectives.

Fracture Critical: Fracture control classification that identifies a part whose individual failure, caused by the presence of a crack, is a catastrophic hazard and that requires safe-life analysis or other fracture control assessment to be shown acceptable for flight. A part is fracture critical unless it can be shown that there is no credible possibility for a flaw to cause failure during its lifetime or the part failure does not result in a credible catastrophic hazard. Assessments for fracture critical parts include damage tolerance analysis, damage tolerance test, or defined approaches for specific categories. Parts under this classification receive flaw screening by NDE, proof test, or process control and are subjected to traceability, materials selection and usage, documentation, and engineering drawing requirements.

Habitable Modules or Volumes: Flight containers/chambers that are designated and designed to support human occupancy.

Hardware Developer: Organization directly responsible for doing the design, manufacture, analysis, test, and safety compliance documentation of the hardware. This includes implementing fracture control requirements.

Hazardous Fluid: For fracture control, a fluid the release of which would create a catastrophic hazard. These types of fluids may include liquid chemical propellants, liquid metals, biohazards, and other highly toxic liquids or gases. The release of such fluids would create a hazardous environment, such as a danger of fire or explosion, unacceptable dilution of breathing oxygen, an increase of oxygen above flammability limits, over-pressurization of a compartment, or loss of a safety-critical system.

Hazardous Fluid Container: Any single, independent (not part of a pressurized system) container or housing that contains a fluid the release of which would cause a catastrophic hazard and that is not classified as a pressure vessel.

Hazardous Material: For fracture control, a material the release of which would create a catastrophic hazard.

High-Cycle Fatigue (HCF): A high-frequency, low-amplitude loading condition created by structural, acoustic, or aerodynamic vibrations that can propagate flaws to failure. An example of an HCF loading condition is the vibrational loading of a turbine blade because of structural resonance.

Impact Damage: The injury or harm inflicted upon composite or bonded hardware by impingement of an object upon the hardware in question or the bumping or striking between the hardware in question and another object. Impact damage is a subset of the more general term damage (or flaw).

Impact Damage Mitigation Plan (IDMP): A plan for composite or bonded hardware to mitigate risk of impact damage to the flight hardware.

Initial Crack (Flaw) Size: The crack size that is assumed to exist at the beginning of a damage tolerance analysis, as determined by NDE or proof testing.

K_c: Plane stress fracture toughness. The value of stress intensity factor K at the tangency between a crack extension resistance curve (R-curve) and the configuration-dependent applied K curve (ASTM E1823, Standard Terminology Relating to Fatigue and Fracture Testing). This crack extension occurs under conditions that do not approach crack-tip plane strain. The R-curve and K_c vary with the material, specimen size, and thickness. K_c is used in NASGRO® to represent fracture toughness as a function of thickness for use in crack growth calculations.[4]

K_{Ic}: Plane strain fracture toughness. The crack extension resistance under conditions of crack-tip plane strain in Mode I for slow rates of loading under predominantly linear-elastic conditions and negligible plastic-zone adjustment that is measured by satisfying a standardized procedure with validity requirements (ASTM E399, Standard Test Method for Linear-Elastic Plane-Strain Fracture Toughness K_{Ic} of Metallic Materials). Another quantity, K_{JIc}, defined for conditions with limited plasticity from J_{Ic} may also be useful (ASTM E1820).

K_{Ie}: Effective fracture toughness for a surface or elliptically shaped crack. The toughness is based on residual strength and the original crack dimensions. This parameter is meaningful only when crack-tip plastic zones are small and stable crack growth before failure is generally absent (ASTM E740/E740M, Standard Practice for Fracture Testing with Surface-Crack Tension Specimens, main body and section X1.2). For conditions with plastic effects and well-defined crack-tip stress fields with fracture controlled by crack initiation, an approach involving constraint may be applicable (ASTM E2899). Testing of flaws in specimens representative of the structure is needed to determine damage tolerance for plasticity conditions when crack-tip stress fields collapse. K_{Ie} is used in NASGRO® for analyses of crack growth.[5]

K_{EAC}: The largest value of the stress intensity factor at which crack growth is not observed for a pre-cracked through-crack specimen of specified material, environment, and thickness that is tested for a significant duration in accordance with ASTM E1681.

K_{IEAC}: The largest value of the stress intensity factor at which crack growth is not observed for a pre-cracked through-crack specimen of specified material, environment, and thickness that is sufficient to meet requirements for plane strain and is tested for a significant duration in accordance with ASTM E1681.

K_{Iscc}: K_{EAC} is often denoted as K_{Iscc} in the literature.

ΔK_{th}: Threshold stress intensity factor range below which flaw growth will not occur under cyclic loading conditions.

---

[4] See NASGRO® User's Manual where the K_c symbol is defined as "critical stress intensity" and section 2.1.4 that shows K_c as a function of material thickness and describes the usage of K_c.

[5] See NASGRO® User's Manual where the K_{Ie} symbol is defined as "effective fracture toughness for part-through (surface/corner) crack" and section 2.1.4 that describes how the K_{Ie} value is determined and how it is used.

<u>Leak-Before-Burst (LBB)</u>: Characteristic of pressurized hardware whose only credible failure mode at or below maximum design pressure (MDP) with service life loads resulting from the presence of a potential flaw is a pressure-relieving leak at the flaw as opposed to burst or rupture at the critical stress intensity factor. As the hardware item leaks down, there is no re-pressurization or continued pressure cycles that could lead to continued crack growth. In this failure mode, the hardware will not fail in a fragmentary, catastrophic manner. Instead, only small, slow-growing leaks would develop, leaking in a controlled manner. Additional aspects of LBB assessments are described in section 6.2.4 in this NASA Technical Standard.

<u>Life Factor</u>: See definition of Service Life Factor.

<u>Lifetime</u>: See definition of Service Life. Refers to a specified life, as opposed to an analytically predicted life.

<u>Limit Load</u>: The maximum load expected on the hardware during its design service life including ground handling, transport to and from orbit, including abort conditions and on-orbit operations.

<u>Limited Life Part</u>: A part that has a predicted damage tolerance life that is less than the required service life factor times the complete service life. See definition of Service Life.

<u>Load Enhancement Factor (LEF)</u>: A factor applied to the service life spectrum to satisfy a specified level of reliability and confidence with fewer cycles than would otherwise be required.

<u>Low-Cycle Loads</u>: A low-frequency, high-amplitude loading condition created by thermal, pressure, or structural loads that can propagate flaws to failure. An example of a low-cycle loading condition is the aerothermal loading of a turbine blade during launch.

<u>Low-Fracture Toughness</u>: Material property characteristic, in the applicable environment, for which the ratio is $K_{Ic}/F_{ty} < 1.66 \sqrt{mm}$ $(0.33 \sqrt{in})$. For steel bolts with unknown K_{Ic}, low-fracture toughness is assumed when material A-basis ultimate strength $F_{tu} > 1,241$ MPa (180 ksi). Parts made with materials of this characteristic may be at risk of a brittle fracture.

<u>Materials Usage Agreement (MUA)</u>: A formal document showing that a noncompliant material is acceptable for the specific application identified.

<u>Maximum Design Pressure</u>: The highest possible operating pressure considering maximum temperature, maximum relief pressure, maximum regulator pressure, and, where applicable, transient pressure excursions. MDP for human-rated hardware is a two-failure tolerant pressure, i.e., it will accommodate any combination of two credible failures that will affect pressure. Some programs have defined MDP as a two-fault tolerant pressure.

Mechanism: A system of moveable and stationary parts that work together as a unit to perform a mechanical function, such as latches, actuators, drive trains, and gimbals.

Mission: A major activity required to accomplish an Agency goal or to effectively pursue a scientific, technological, or engineering opportunity directly related to an Agency goal. Mission needs are independent of any particular system or technological solution (NPR 7120.5, NASA Space Flight Program and Project Management Requirements).

Net-Section Stress or Strain: The stresses or strains computed for a hypothetical cut across a part, based on strength-of-materials theory. Possible bending loads can produce stress gradients across the net section, in which case the net-section stress is found to be the maximum combination of tension and bending stress, ignoring geometric stress concentrations. (An example of net-section stress calculation detailed in the NASGRO® User's Manual, Appendix B.)

No-Growth Threshold Strain: For a composite or bonded part, the largest strain range (where strain range is the maximum absolute value of strain in a load cycle) below which flaws compatible with the sizes established by NDE, special visual inspection, the DTA, or the minimum sizes imposed do not grow in 10^6 cycles (10^8 cycles for rotating hardware) at a load ratio appropriate to the application. Thresholds are determined on specimens with flaws for which sufficient load/cycles have been initially applied to cause flaw growth. The no-growth threshold strain is a function of the material and layup and is determined from test data in the appropriate environment for the applicable (or worst) orientation of strain and flaw for a particular design.

Non-Destructive Evaluation: Examination of parts for flaws using established and standardized inspection techniques that are harmless to hardware, such as radiography, penetrant, ultrasonic, magnetic particle, and eddy current. NDE is sometimes referred to as non-destructive testing (NDT) or non-destructive inspection (NDI).

Non-Hazardous-Leak-Before-Burst (NHLBB): A non-fracture critical classification for metallic pressurized hardware that contains a material that is not hazardous and that exhibits the LBB failure mode in a non-hazardous manner.

Part: Hardware item considered a single entity for the purpose of fracture control.

Pressure Vessel: A container designed primarily for pressurized storage of gases or liquids and that also performs any of the following:

a. Contains stored energy of 19,307 J (14,240 ft-lb) or greater based on adiabatic expansion of a perfect gas.

b. Stores a gas that will experience an MDP greater than 690 kPa (100 psia).

c. Contains a gas or liquid in excess of 103 kPa (15 psia) that will create a catastrophic hazard if released.

Pressurized Component: A line, fitting, valve, regulator, etc., that is part of a pressurized system intended primarily to sustain a fluid pressure and fluid transfer. Any piece of hardware that is not a pressure vessel or a pressurized fluid container but is pressurized via a pressurization system.

Pressurized Fluid Container: A container designed primarily for pressurized storage of gases or liquids that is similar to a pressure vessel but does not satisfy the definition of a pressure vessel.

Pressurized Hardware: Any of the various hardware items that support an internal pressure.

Pressurized Structure: A hardware item designed to carry both internal pressure and vehicle structural load.

Pressurized System: An interrelated configuration of pressurized components under positive internal pressure. The system may also include pressure vessels.

Proof Test: A test on the flight article that is performed to verify structural acceptability or to screen flaws. The proof test load and/or pressure level is the proof test factor times limit load and/or MDP. Proof tests may be conducted in the operational environment, or the test levels may be adjusted via an ECF. (Note that some sections within this NASA Technical Standard may specify when an ECF is optional versus when it is prescribed for the classification if the test is not conducted in the operational environment.)

Proof Test Factor: A factor that is multiplied by the limit load and/or MDP to arrive at the proof test levels. When proof tests are performed to establish structural acceptability, the proof test factor is specified. When screening for flaws with a proof test, the proof test factor is derived by fracture mechanics principles.

R Ratio: The ratio of minimum stress to maximum stress in cyclic loading.

Re-flight Hardware: Hardware items that have already met the requirements in this NASA Technical Standard for service life, have flown on a flight vehicle, and are being manifested for an additional flight. Note that some fracture control categories in this NASA Technical Standard impose additional requirements that are to be satisfied before being re-flown.

Residual Strength: The maximum value of load (both externally applied and internal self-equilibrating loading, such as residual stresses) that a flawed or damaged part is capable of sustaining without catastrophic failure.[6]

Residual Threat Determination: An assessment that defines the worst-case credible flaw conditions that composite or bonded hardware will be designed to endure, considering all applicable flaw detection and mitigation strategies that are implemented for the flight hardware.

Responsible NASA Center: The NASA Center acting as the sponsor and/or coordinator for the program/project developing the payload/hardware.

Responsible Fracture Control Board: A designated multi-discipline group of experts at the NASA Center that has the authority to develop, interpret, and approve fracture control requirements and the responsibility for overseeing and approving the technical adequacy of all fracture control activities at the Center.

Rotating Hardware: Hardware that has a rotational mode of operation and devices with spinning parts, such as fans, centrifuges, motors, pumps, gyros, and flywheels.

Rupture: An instance of breaking or bursting suddenly and completely.

Safe Life: See definition of Damage Tolerance.

Safety Critical: For fracture control, a part, component, or system whose failure or loss would be a catastrophic hazard.

Sealed Container: Any single, independent container (not part of a pressurized system), component, or housing that is sealed to maintain an internal non-hazardous environment and that does not meet the definition of a pressure vessel.

Service Life: Time interval for a part beginning with manufacture and extending throughout all phases of its specified mission usage. The period of time or number of cycles that includes all relevant loadings, conditions, and environments encountered during this period that will affect flaw growth, including all manufacturing, testing, storage, transportation, launch, on-orbit, descent, landing, and if applicable, post-landing events, refurbishments, retesting, and repeated flights until the hardware is retired from service.

Service Life Factor: The factor on service life required in damage tolerance analysis or testing. The service life factor is often referred to as the life factor. (Note that the service life factor is specified as 4 for metallic materials in section 7.3.2.c in this NASA Technical Standard. The service life factor is specified as the B-basis number of service lives with the corresponding LEF for composites or bonded materials in sections 7.4.7.b and 7.4.8.e in this NASA Technical Standard.)

---

[6] In the NASGRO® User Manual version 7.1.1, section 2.1.5 and Appendix O, there is discussion of a related failure condition invoked when net section stress exceeds the material flow stress.

Shatterable Materials: Any material that is prone to brittle failures during operation that could release many small pieces into the surrounding environment.

Special Visual Inspection: Close proximity, intense visual examination of localized areas of internal and/or external structure for indications of impact damage, flaws, or other structural anomaly. Appropriate access to gain proximity, e.g., removal of fairings and access doors, use of ladders and work stands, is required. High-intensity lighting, along with other inspection aids such as mirrors, magnifying lenses, and surface cleaning, are used. Special visual inspections are done independently by two inspectors. When special visual indications are found, NDE is done.

Standard NDE: NDE methods of metallic materials for which a statistically based flaw detection capability has been established. Standard NDE methods addressed by this document are limited to fluorescent penetrant, radiography, ultrasonic, eddy current, and magnetic particle.

Sustained Load Cracking (SLC): Growth of a pre-existing crack in susceptible metallic alloys[7] under sustained stress without assistance from an external environment. A threshold stress intensity factor can be obtained by procedures such as those in ASTM E1681 for the case of an inert or vacuum environment. One publication determines the effects of hydrogen content and temperature on SLC in Ti-6Al-4V (Boyer and Spurr, 1978).

Ultimate Factor of Safety (Ultimate Safety Factor): A specified factor to be applied to limit load. No ultimate structural failure is allowed for a load equal to the ultimate factor of safety multiplied times limit load.

Ultimate Strength (Capability): The load, stress, or strain at which collapse or rupture occurs.

Yield Strength: The stress that corresponds to a plastic axial strain of 0.002 mm/mm (0.002 in/in).

---

[7] SLC, because of the presence of interstitial hydrogen, occurs in titanium alloys, including commercially pure titanium (cp-Ti) and Ti-6Al-4V (Ti64), in both annealed and solution treated and aged (STA) conditions. Testing is necessary to determine the threshold stress intensity for the titanium alloy metallurgical condition and interstitial hydrogen content. Other materials with different crystalline structures such as steel and aluminum alloys that do not allow interstitial hydrogen may still exhibit SLC behaviors.

4. GENERAL REQUIREMENTS

4.1 Fracture Control Plan

A summary table of all FCRs in this NASA Technical Standard is shown in Appendix B in this NASA Technical Standard.

[FCR 1] A Fracture Control Plan shall be developed and maintained by the program for human-rated spaceflight hardware that satisfies all of the following:

    a. Addresses all of the parts in the program-specific flight hardware.

    b. Meets the requirements of this NASA Technical Standard.

    c. Specifies fracture controls that are established to mitigate the risk of catastrophic failure caused by flaws throughout the service life of the hardware.

    d. Has approval by the RFCB.

[Rationale: The FCP is necessary to document the hardware-specific fracture control requirements, such as parts classification, selected approaches for each part, and all required fracture control activities for the program or project. The RFCB-approved FCP is the working document that all responsible parties use for implementing fracture control requirements to a particular program or project.]

The FCP details the fracture control responsibilities, the classification of all parts for the specific hardware, the selected applicable fracture control approaches from the requirements of this NASA Technical Standard corresponding to each part's category, as well as the approaches for flaw screening, traceability, and material selection of fracture critical parts. The hardware-specific FCP also documents all alternative approaches in accordance with the requirement of [FCR 26] in this NASA Technical Standard.

Each separate hardware project within a program may develop an FCP for its hardware.

The initial FCP should be submitted early in the program. An early draft and subsequent updates of an FCP are necessary for appropriate cost estimation. The Data Requirements Deliverable for the FCP may need to be updated to require an earlier draft (potentially as part of the hardware proposal) for more accurate cost estimation.

The FCP should be updated as needed to keep it current with the documented program fracture control approaches.

4.2   Responsibilities

4.2.1   Responsible Fracture Control Board

[FCR 2] The NASA Center responsible for the human-rated spaceflight hardware shall establish and designate a NASA RFCB to ensure compliance with the technical requirements of this document.

[Rationale: The purpose of this requirement is to clearly establish a NASA RFCB as the body responsible for assuring technical compliance with this NASA Technical Standard.]

4.2.2   Responsible Program

[FCR 3] Human-rated spaceflight programs shall impose fracture control on their projects to meet the requirements of this NASA Technical Standard.

[Rationale: The purpose of this requirement is to ensure this NASA Technical Standard is applied to all human-rated spaceflight programs, including those for unmanned vehicles that approach or dock with human-rated vehicles, such as the International Space Station or Orion. Implementation of fracture control mitigates the risk of catastrophic structural failure related to flaws, thereby increasing reliability of the hardware and the safety of the crew.]

4.2.3   Fracture Control Implementation

[FCR 4] Fracture control implementation shall be performed with the oversight, advice, and approval of the RFCB.

[Rationale: This requirement identifies the RFCB as the technical body responsible for determining the adequacy of fracture control implementation. This determination includes assessing whether the project is deploying sufficient technical capabilities and processes for fracture control. It includes monitoring of damage tolerance assessments and hardware verification activities to assure that all hardware complies with the requirements in this document. To accomplish those goals, the RFCB should have opportunities to review and comment on these activities and have access to all the technical information needed to confirm compliance with this document.]

Each project should identify organizational elements (or technical disciplines) and their responsibilities for implementing and documenting fracture control aspects that affect hardware design, manufacturing, inspections, and planned operations. These responsibilities should be identified at project formulation and documented in the FCP. The organizational elements that implement fracture control and assess current hardware developments should be part of project milestone reviews associated with structural integrity and safety. The RFCB should have an opportunity to participate in and receive summaries of major project reviews as the program formulates system and hardware requirements, as well as when the hardware developer designs and selects technical approaches for meeting fracture control requirements, generates hardware fracture control assessments, and reports on relevant testing.

4.3   Classification of Parts and Implementation Requirements

[FCR 5] All parts used in human-rated spaceflight hardware shall be evaluated to identify the following:

      a.   The fracture control classification of each part as either exempt, NFC, or fracture critical.

      b.   The corresponding approaches that follow the requirements of this NASA Technical Standard to be documented in the FCP.

[Rationale: To implement fracture control, all parts need to be evaluated for criticality to assure they are appropriately classified and to identify subsequent activities related to the classification. Not all parts are fracture critical.]

The approach implemented for fracture control classification of parts is documented in the FCP as described in section 4.1 [FCR 1].

All parts go through a fracture control classification process for all mission phases to determine which parts are fracture critical. Parts may be classified as one of the following:

      a. Exempt.
      b. NFC.
      c. Fracture critical.

Parts classified as exempt are to be exempt for all phases of the service life of the part. Parts that are fracture critical in any phase of the service life of the part are classified as fracture critical. Parts that do not fit into the exempt or NFC categories are to be classified and evaluated as fracture critical parts.

Approaches to evaluate hardware in these three categories are presented in sections 5, 6, and 7 in this NASA Technical Standard. Figure 2, NASA-STD-5019A Fracture Control Classification Logic Diagram, shows a logic diagram for the classification of parts and references to the applicable sections of this document. Figure 3, Fracture Control Assessment Process and Activities Corresponding to Parts Classifications, is a chart with a general description of activities associated with each classification.

If hardware that was certified to earlier fracture control requirements levied under earlier programs is to be flown under a new program, then the hardware should be re-assessed using the fracture control requirements of this document. Additionally, hardware that experiences service life conditions that deviate from the certified design configuration or conditions, either through off-nominal service conditions or degradation during service, is to be re-assessed in accordance with the requirements of this document.

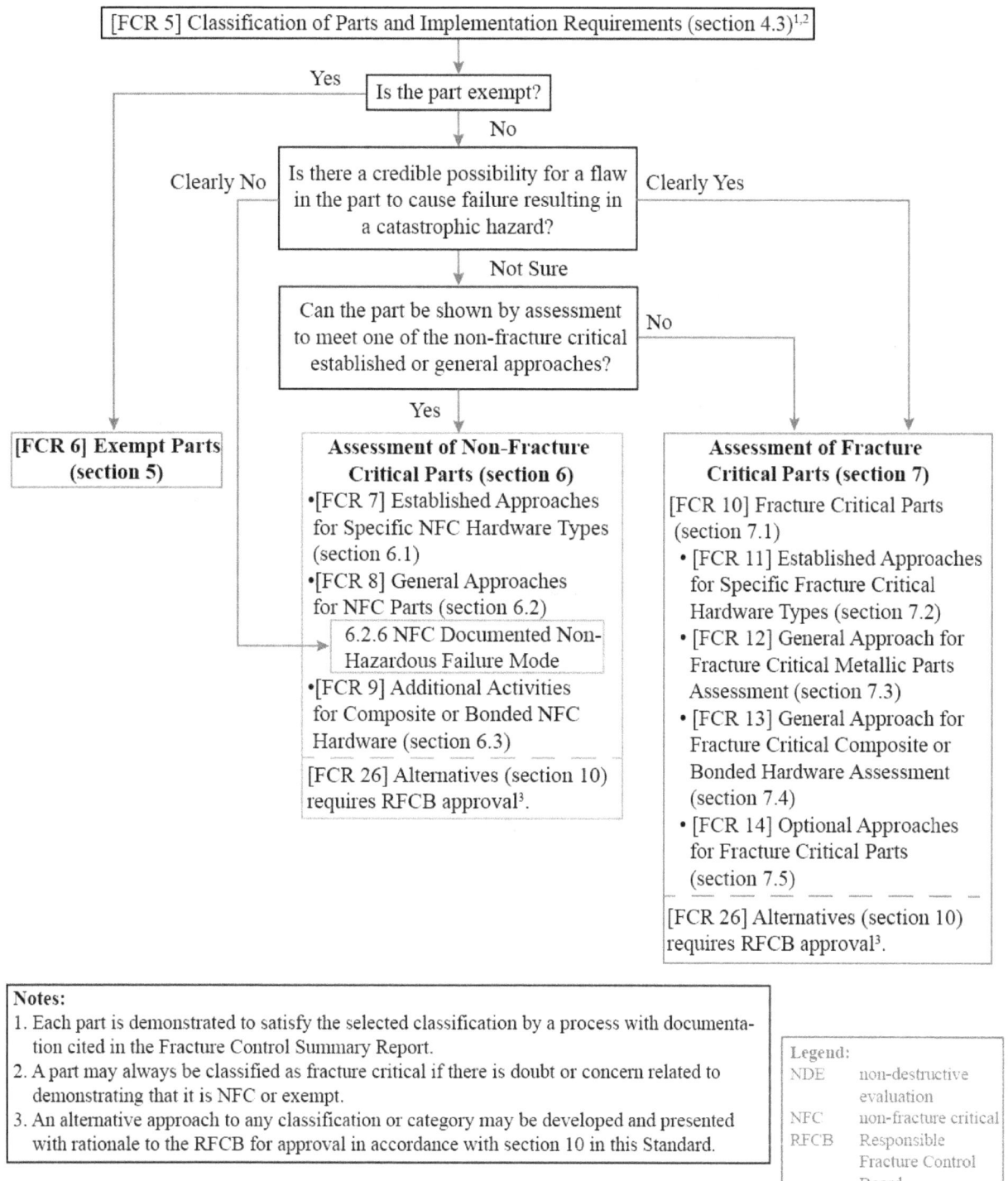

Figure 2—NASA-STD-5019A Fracture Control Classification Logic Diagram

Classification	[FCR 6] Exempt Parts (section 5)	Non-Fracture Critical Parts (section 6)	[FCR 10] Fracture Critical Parts (section 7)
Approach	Parts fit in one of the following: •[FCR 6] a. Non-structural parts with no credible failure mode caused by a flaw •[FCR 6] b. Non-structural parts with no credible potential for causing a catastrophic hazard •[FCR 6] c. Other non-structural parts approved by the RFCB for exempt status	**Assessment of NFC Parts (section 6)** •[FCR 7] Established Approaches for Specific NFC Hardware Types (section 6.1) •[FCR 8] General Approaches for NFC Parts (section 6.2) •[FCR 9] Additional Activities for Composite or Bonded NFC Hardware (section 6.3) [FCR 26] Alternatives (section 10) requires RFCB approval.	**Assessment of Fracture Critical Parts (section 7)** [FCR 10] Fracture Critical Parts (section 7.1) • [FCR 11] Established Approaches for Specific Fracture Critical Hardware Types (section 7.2) • [FCR 12] General Approach for Fracture Critical Metallic Parts Assessment (section 7.3) • [FCR 13] General Approach for Fracture Critical Composite or Bonded Hardware Assessment (section 7.4) • [FCR 14] Optional Approaches for Fracture Critical Parts (section 7.5) [FCR 26] Alternatives (section 10) requires RFCB approval.
Actions Required*	No additional action beyond the FCP and FCSR documentation	Various, including analysis, test, inspection, and verification. Unique to NFC Composite or Bonded: •Damage Threat Assessment •Impact Damage Mitigation Plan •Residual Threat Determination •Flaw Screening with NDE •Material Selection •Traceability	•Damage Tolerance Analysis/Test •Flaw Screening •Traceability •Material Selection •Verification Unique to Composite or Bonded: •Damage Threat Assessment •Impact Damage Mitigation Plan •Residual Threat Determination •Building Block Test and Analysis

*Note: Documentation is required for **all** classifications [FCR 1]. In addition to these unique actions, documentation of the approach used to demonstrate that each part satisfies the selected classification is to be cited in the Fracture Control Summary Report [FCR 24].

Legend:	
FCP	Fracture Control Plan
FCSR	Fracture Control Summary Report
NDE	non-destructive evaluation
NFC	non-fracture critical
RFCB	Responsible Fracture Control Board

Figure 3—Fracture Control Assessment Process and Activities Corresponding to Parts Classifications

4.4 Other Requirements

Implementation of fracture control and full compliance with fracture control requirements do not relieve the hardware developer from compliance with other hardware design and test requirements, quality assurance requirements, or materials requirements that are applicable independent of fracture control. The hardware developer should be aware of the influence of other requirements on the implementation of fracture control activities. For example, NASA-STD-5001, Structural Design and Test Factors of Safety for Spaceflight Hardware, requires that all composite or bonded flight hardware be acceptance (proof) tested, whereas this NASA Technical Standard does not require such tests in general.

5. EXEMPT PARTS

In some cases, parts may be classified as exempt.

[FCR 6] Each part classified as exempt shall fit into one of the following categories:

   a. Non-structural parts with no credible failure mode caused by a flaw.
   b. Non-structural parts with no credible potential for causing a catastrophic hazard.
   c. Other non-structural parts approved by the RFCB for exempt status.

[Rationale: Non-structural parts that do not have a credible failure mode caused by a flaw and those with no credible potential for causing a catastrophic hazard do not need to be assessed for fracture criticality because they do not pose a catastrophic hazard.]

Use of an alternative approach requires unique rationale and approval by the RFCB as described in section 10 [FCR 26] in this NASA Technical Standard.

Parts that are identified and shown to meet the exempt classification criteria in documentation cited in the Fracture Control Summary Report (FCSR) in accordance with the requirements listed in section 9 of this NASA Technical Standard comply with fracture control requirements without further activity beyond conventional aerospace verification and quality assurance procedures, unless otherwise indicated in this document.

Exempt parts typically include non-structural items or items that do not have a credible failure mode related to the presence of a flaw, such as flexible insulation blankets, enclosed electrical circuit components/boards, wire bundles, and certain batteries listed in section 6.1.6 in this NASA Technical Standard. The RFCB may accept other items as exempt based on rigorous development programs that establish their safety and functional reliability.

Non-structural parts are generally those not intended to resist loads. A part that may be structural in one system or application may not be in another. Discussion with the RFCB may be necessary.

6. ASSESSMENT OF NON-FRACTURE CRITICAL PARTS

In some cases, parts may be classified as NFC. During the classification of all parts in accordance with section 4.3 [FCR 5] in this NASA Technical Standard, the hardware developer identifies the applicable approaches or activities of this section for each of the parts that are to be classified as NFC and documents them in their hardware-specific FCP. The RFCB reviews and approves the FCP to ensure that it includes all specified elements of the selected approaches.

The methods in this section are based on NASA's experience base, established approaches, industry standards, or aerospace standards. Any deviations or omissions of elements in the activities or approaches prescribed in this section constitute an alternative approach that is to satisfy the requirements in section 10 [FCR 26] in this NASA Technical Standard.

NFC parts are those shown in the hardware-specific FCP to meet a category in this section, which typically involves documentation of an assessment involving some combination of analysis, test, inspection, failure mode evaluation, or adherence to specified criteria listed in each subcategory. The documentation is to be cited in the FCSR where the part is listed as NFC in accordance with section 9 requirements in this NASA Technical Standard.

Parts that are identified and shown to meet NFC classification criteria in documentation cited in the FCSR in accordance with section 9 requirements in this NASA Technical Standard comply with fracture control requirements without further activity beyond conventional aerospace verification and quality assurance procedures, unless otherwise indicated in this document.

6.1 Established Approaches for Specific NFC Hardware Types

Parts in this category are classified NFC if documented assessment cited in the FCSR shows they satisfy the specified criteria listed in the item corresponding to the hardware type. Composite and bonded hardware are to satisfy section 6.3 in this NASA Technical Standard in addition to requirements for a specific hardware type.

[FCR 7] To be classified as NFC, each part that is described by a specific hardware type in the following list shall comply with the established approach given in the referenced subsection:

   a.   NFC metallic fasteners, rivets, shear pins, and locking devices comply with section 6.1.1 in this NASA Technical Standard.

   b.   NFC shatterable components and structures comply with section 6.1.2 in this NASA Technical Standard.

   c.   NFC rotating hardware complies with section 6.1.3 in this NASA Technical Standard.

   d.   NFC sealed containers comply with section 6.1.4 in this NASA Technical Standard.

e. NFC tools, mechanisms, and tethers comply with section 6.1.5 in this NASA Technical Standard.

f. NFC batteries comply with section 6.1.6 in this NASA Technical Standard.

[Rationale: Parts that can be shown to have no credible catastrophic hazard resulting from a failure of the part caused by a flaw or to have no credible possibility for flaws to cause failure are not fracture critical. These parts can be classified as NFC. To assist this classification process, a number of established approaches have been developed for specific hardware types that are documented in this NASA Technical Standard.]

Use of an alternative approach requires unique rationale and approval by the RFCB as described in section 10 [FCR 26] in this NASA Technical Standard.

6.1.1 NFC Metallic Fasteners, Rivets, Shear Pins, and Locking Devices

To classify a part as an NFC metallic fastener, rivet, shear pin, or locking device, satisfy any of the following items in sections 6.1.1.1 through 6.1.1.6, depending on application, hardware type, and failure modes, to comply with requirement [FCR 7], section 6.1.a in this NASA Technical Standard.

6.1.1.1 NFC Low-Released Mass Fasteners, Rivets, and Shear Pins

To classify a metallic fastener, rivet, or pin as NFC low-released mass, meet the requirements of section 6.2.1 in this NASA Technical Standard.

A metallic fastener, rivet, or pin that has an individual single-point structural failure or a group of fasteners, rivets, or pins where loss of any one fastener, rivet, or pin does not present a catastrophic hazard can be classified in one of these NFC categories.

6.1.1.2 NFC Contained Fasteners, Rivets, and Shear Pins

To classify a metallic fastener, rivet, or pin as NFC contained, meet the requirements of section 6.2.2 in this NASA Technical Standard.

A metallic fastener, rivet, or pin that has an individual single-point structural failure or a group of fasteners, rivets, or pins where loss of any one fastener, rivet, or pin does not present a catastrophic hazard can be classified in one of these NFC categories.

6.1.1.3 NFC Fail-Safe Rivets

To classify metallic rivet applications as NFC fail-safe, meet the requirements in section 6.2.3 in this NASA Technical Standard.

6.1.1.4 NFC Low-Risk Fasteners

To classify metallic fasteners as NFC low risk, meet the following:

a. The fastener is in a local pattern of two or more similar fasteners.

b. The fastener satisfies all of the following general fastener attributes:

(1) Fasteners are fabricated from a metal with high resistance to stress corrosion cracking, as defined in MSFC-STD-3029, Guidelines for the Selection of Metallic Materials for Stress Corrosion Cracking Resistance in Sodium Chloride Environments.

(2) Fasteners are fabricated to a military, NAS, or commercial aerospace specification approved by the procuring organization.

    A. The standard and/or associated procurement specification includes tensile, shear, and fatigue testing as part of lot acceptance.

    B. Fasteners with complete traceability are delivered with the Material Test Report or equivalent that includes the following:

        i. The raw material and heat-treat certifications.

        ii. Documentation of applicable testing or processing required in the associated procurement specification.

Fasteners that are manufactured from the following list of ductile materials show a high tolerance for typical fastener defects and flaws. These are typically accepted as low-risk fasteners. Examples of procurement specifications for these commonly accepted low-risk fastener materials are:

- *Iron-based superalloy A286: NAS4003, Fastener, A286 Corrosion Resistant Alloy, Externally Threaded, 160 KSI F_{tu}, 95 KSI F_{su}, 1000 °F; NA0026, Procurement Specification Metric Fasteners, A-286 CRES Externally Threaded, 1100 MPa Tensile, 660 MPa Shear; or equivalent.*
- *Nickel-based superalloy Inconel 718: NASM85604, Bolt, Nickel Alloy 718, Tension, High Strength, 125 KSI F_{su} and 220 KSI F_{tu}, High Temperature, Spline Drive; or equivalent.*
- *Cobalt-Chromium-Nickel-based superalloy MP35N: AS7468, Bolts, Cobalt-Chromium-Nickel Alloy, UNS R30035, Tensile Strength 260 Ksi, Procurement Specification; or equivalent.*

- *Austenitic Stainless Steel 300 Series CRES: NA0271, Metric Fasteners, CRES 300 Series, Externally Threaded, MJ Thread, 500 MPa F_{tu} and 700 MPa F_{tu}; or equivalent.*

(3) Fasteners are not made from any titanium alloy.

Titanium alloys, such as Ti-6Al-4V, cp-Ti, and other titanium alloys are not acceptable in this category because of generic EAC or SLC failure modes, as well as low fracture toughness.

(4) The fastened joint complies with both of the following from NASA-STD-5020, Requirements for Threaded Fastening Systems in Spaceflight Hardware:

    A. Preload control as detailed in section 6.1 of NASA-STD-5020.

    B. No joint separation in the nominal loading configuration as described in sections 4.3 and 6.5 of NASA-STD-5020.

(5) Fasteners are subject to the following:

    A. Have rolled threads, with the rolling process occurring after all thermal treatment of the material.

    B. The results of the mandatory lot acceptance fatigue testing are required to establish that the fasteners meet the fatigue requirements in NASA-STD-5001.

    C. For fastener types that do not require fatigue testing as part of lot acceptance, samples from the lot need to be submitted for fatigue testing in accordance with NASM1312-11, Fastener Test Methods, Method 11 Tension Fatigue, and NAM1312-111, Fastener Test Methods, Method 111 Tension Fatigue, or equivalent, to satisfy 6.1.1.4.b.(2) above.

(6) The fasteners are not made from a low fracture toughness alloy, as defined in section 3.2 in this NASA Technical Standard.

(7) Fasteners are not reworked or custom made unless the application is approved by the RFCB.

6.1.1.5   NFC Fail-Safe Fasteners

To classify metallic fasteners as NFC fail-safe, meet the following:

    a. The fasteners meet the fail-safe requirements in section 6.2.3 in this NASA Technical Standard.

b. The fastener satisfies all of the following general fastener attributes:

(1) Fasteners are fabricated from a metal with high resistance to stress corrosion cracking as defined in MSFC-STD-3029.

(2) Fasteners are fabricated, procured, and inspected in accordance with NASA-STD-6008, NASA Fastener Procurement, Receiving Inspection, and Storage Practices for Spaceflight Hardware, and an equivalent military standard, NAS, proprietary, or commercial aerospace specification approved by the procuring organization.

(3) The fastened joint complies with NASA-STD-5020 without joint separation in the nominal configuration.

(4) Fasteners have rolled threads and are assessed to establish that they meet the fatigue requirements in NASA-STD-5001.

(5) The fasteners are not made from a low fracture toughness alloy as defined in section 3.2 in this NASA Technical Standard.

(6) Fasteners are not reworked or custom made unless the application is approved by the RFCB.

(7) Fasteners manufactured from titanium alloys require additional considerations for this classification, including risk mitigation and assessment that are approved by the RFCB.

Titanium alloys, such as Ti-6Al-4V (including annealed and STA conditions), cp-Ti, and other titanium alloys, have potential generic EAC or SLC failure modes that are not mitigated by the fail-safe requirements. Additional risk mitigation is needed for their use in this classification with an assessment that establishes that there is no credible risk of generic fastener failures related to flaws or under applied load. The assessment should include credible initial fastener defects/crack size, the largest credible preload, and maximum service life loading and should compare the Critical Stress Intensity Factor to K_{SLC} and K_{EAC} lower bound values determined from tests of flawed fasteners in applicable service life environments.

6.1.1.6   NFC Locking Devices

To classify metallic locking devices to prevent fastener or connector backout, including wires, tangs, or other methods, as NFC locking devices, NFC hardware is to meet the requirements of section 6.2.1 in this NASA Technical Standard.

6.1.2 NFC Shatterable Components and Structures

To classify parts as NFC shatterable components or structures, satisfy one of the items listed below in section 6.1.2.1 or section 6.1.2.2 in this NASA Technical Standard, depending on application and hardware type, to meet requirement [FCR 7], section 6.1.b in this NASA Technical Standard.

6.1.2.1 NFC Internal Shatterable Components

Shatterable components and structures are classified as NFC by one of the following:

    a. For shatterable components and structures inside habitable volumes, meet the following:

        (1) Requirements in section 6.2.2 in this NASA Technical Standard.

        (2) The particulate containment requirements in NASA-STD-5018, Strength Design and Verification Criteria for Glass, Ceramics, and Windows in Human Space Flight Applications.

    b. For shatterable components and structures inside non-habitable volumes, meet the requirements in sections 6.2.1, 6.2.2, 6.2.3, 6.2.4, or 6.2.6 in this NASA Technical Standard.

6.1.2.2 NFC External Shatterable Components

To classify parts as NFC shatterable components or structures located on the external surface of a spacecraft that are manufactured from a material with limited ductility such that it is prone to brittle failures when cracked and/or subjected to impact, meet 6.1.2.2.a, 6.1.2.2.b, and 6.1.2.2.c below:

    a. A DTA and IDMP are developed to mitigate credible catastrophic impacts from vehicle loss of external surface mass, crew exposure, micrometeoroid and orbital debris (MMOD), extravehicular activity (EVA), inadvertent contacts, and EVA tool impact hazards.

    b. The design has sufficient structural integrity such that the loss of a primary member does not result in catastrophic loss of spaceflight hardware function or required strength that prevents the hardware from safely completing the mission.

    c. Any mass released from these components meets the low-released mass requirements of section 6.2.1 in this NASA Technical Standard.

Refer to section 7.4.1 for DTA and 7.4.2 for IDMP.

6.1.3 NFC Rotating Hardware

Satisfy one of the following items to classify a part as NFC rotating hardware to meet requirement [FCR 7], section 6.1.c in this NASA Technical Standard:

a.   The rotating hardware is computer equipment, such as computer data storage disks and computer cooling fans.

b.   The rotating hardware meets the conditions in section 6.2.5 in this NASA Technical Standard.

c.   The rotating hardware is within an enclosure and meets the following:

 (1) In the event of a rotor fracture caused by flaws, a conservative assessment of credible rotor fragments shows the fragments are contained within the enclosure in accordance with section 6.2.2 in this NASA Technical Standard.

 (2) The structural mounts for the rotating hardware and the enclosure are evaluated as standard structure and meet fracture control requirements.

 (3) The mount assessments include credible loads from a sudden stop of the rotor, unless it is established that either of the following are satisfied:

 A.   The rotating hardware does not have a credible sudden stop catastrophic hazard during the service life that has resulted from a structural failure of the rotating hardware or adjacent structure caused by flaws.

 B.   The rotating hardware has design features and monitoring with safety controls that make a sudden stop a non-credible event.

6.1.4 NFC Sealed Containers

Satisfy all of the following to classify a part as an NFC sealed container, e.g., a sealed electronics box that is not part of a pressure system and is not a pressure vessel, to meet requirement [FCR 7] section 6.1.d in this NASA Technical Standard:

a.   The container meets the following:

 (1) Does not contain a hazardous material.

 (2) Loss of pressure or fluid from the container does not result in a catastrophic hazard.

 (3) Container supports meet fracture control requirements.

Note that supports may either be integral to the container or separate parts, such as brackets. If the supports are integral to the container and are fracture critical, further discussion with the RFCB is necessary for classification of the container. Separate support parts should be classified independent of the container.

b.   The part is manufactured from metal alloys typically used for commercially available sealed containers procured to an aerospace standard or equivalent that are not susceptible to crack extension related to EAC or SLC.

c.   The container satisfies the LBB definition in this document at MDP.

d.   The container does not have an impervious barrier or coating that inhibits leakage on either the interior or exterior surfaces.

e.   A container is subject to the following:

(1) Inspected for leaks before repressurization.

(2) Re–flight containers are inspected for leaks before being re-flown.

f.   The container stored fluid energy is less than 19,307 J (14,240 ft-lb) based on adiabatic expansion of a perfect gas.

g.   If the MDP of the container is 152 kPa (22 psi, 1.5 atm) or less, no additional assessment for items h and i below is required.

h.   If the MDP of the container is greater than 152 kPa (22 psi, 1.5 atm) and no more than 304 kPa (44 psi, 3 atm), satisfy one of the following:

(1) An analysis shows that the container has a positive margin against burst when a factor of 2.5 on MDP is used.

(2) The container is proof tested to a minimum of 1.5 times the MDP.

i.   If MDP is greater than 304 kPa (44 psi, 3 atm), the sealed container may not be classified in this category.

The container portion of an NFC sealed container does not require NDE to screen for flaws. The container supports may require NDE, depending on their individual fracture control classification.

The guidance on LBB assessment provided in section 6.2.4 in this NASA Technical Standard is also applicable to this section. Proof tests are usually performed in the operational environment, or the test levels are adjusted via an ECF.

Inertial load effects (including attach points) may necessitate additional assessments beyond the items in this category.

6.1.5   NFC Tools, Mechanisms, and Tethers

Satisfy either of the following to classify a part as an NFC tool, mechanism, or tether to meet requirement [FCR 7] section 6.1.e in this NASA Technical Standard:

a.   To classify tools, mechanisms, and tethers as NFC during storage and usage, meet the requirements in sections 6.2.1 or 6.2.5 in this NASA Technical Standard.

b.   To classify tools, mechanisms, and tethers as NFC during storage, meet the requirements in sections 6.2.2 or 6.2.3 in this NASA Technical Standard.

6.1.6   NFC Batteries

Satisfy one of the following to classify parts as NFC battery cells/cases to meet requirement [FCR 7] section 6.1.f in this NASA Technical Standard:

a.   Meet the NHLBB requirements in section 6.2.4 in this NASA Technical Standard.

b.   Meet the sealed container requirements in section 6.1.4 in this NASA Technical Standard.

Small batteries in common use, such as button cells of 200 mA-hr or less and carbon-zinc or zinc-air batteries of size F or smaller are exempt from fracture control.

6.2   General Approaches for NFC Parts

Parts in this category may be classified NFC if documented assessment cited in the FCSR shows that they do not present a credible catastrophic hazard resulting from failure of the part caused by a flaw or that they do not have a credible possibility for a flaw to cause failure of the part. Both composite and bonded hardware are to satisfy section 6.3 in addition to the items in this section.

[FCR 8] Each part classified as NFC that is not of a specific hardware type as described in section 6.1 in this NASA Technical Standard shall comply with one of the following items:

a.   NFC low-released mass complies with section 6.2.1 in this NASA Technical Standard.

b.   NFC contained complies with section 6.2.2 in this NASA Technical Standard.

c.   NFC fail-safe complies with section 6.2.3 in this NASA Technical Standard.

    d. NFC NHLBB pressurized components comply with section 6.2.4 in this NASA Technical Standard.

    e. NFC low-risk part complies with section 6.2.5 in this NASA Technical Standard.

    f. NFC documented non-hazardous failure mode complies with section 6.2.6 in this NASA Technical Standard.

[Rationale: Parts that can be shown to have no credible catastrophic hazard resulting from a failure of the part caused by a flaw or to have no credible possibility for flaws to cause failure are not fracture critical. These parts can be classified as NFC.]

Use of an alternative approach requires unique rationale and approval by the RFCB as described in section 10 [FCR 26] in this NASA Technical Standard.

Sections 6.2.1 through 6.2.4 in this NASA Technical Standard provide approaches to establish that a part does not present a credible catastrophic hazard because of part failure. Section 6.2.5 in this NASA Technical Standard provides an approach to show that a part does not have a credible possibility for a flaw to cause failure in the part.

6.2.1 NFC Low-Released Mass

Small parts or masses that are released (because of structural failure caused by a flaw) may be designated NFC via the low-released mass category.

Satisfy all of the following items to classify a part as an NFC low-released mass to meet requirement [FCR 8] section 6.2.a in this NASA Technical Standard.

    a. The fracture of the part does not cause a catastrophic hazard.

    b. The release of the mass does not cause a catastrophic hazard.

        (1) For NFC composite or bonded parts that may be impacted by an NFC low-released mass part, establish that the impacted NFC composite or bonded parts can sustain DUL. This is verified by analysis combined with coupon or hardware element test data while subject to the worst-case impact damage from the released mass.

        (2) For fracture critical composite or bonded parts that may be impacted by an NFC low-released mass part, include the worst-case impact damage from the released mass in the DTA and RTD during evaluation of the fracture critical part, described in section 7.4 in this NASA Technical Standard.

        (3) Loss of function and impact with other hardware, equipment, spacecraft, and personnel are addressed in the evaluation.

(4) External released mass or parts, including those that would be subjected to aerodynamic flow, may only be classified low-released mass when the program has established an acceptable debris field criterion and the parts fall within it.

The program should provide the launch vehicle acceptable debris field criteria. The program or launch payload integrator has to address concerns of impact on adjacent payloads and other spacecraft.

6.2.2 NFC Contained

Parts that would be safely confined to an enclosed volume should they become loose because of the presence of a flaw may be designated NFC via the contained category.

Satisfy all of the following items to classify a part as NFC contained to meet requirement [FCR 8] section 6.2.b in this NASA Technical Standard:

    a. A containment assessment conservatively establishes that the contained part does not penetrate, fracture, or otherwise escape the enclosure.

        (1) Metallic containers are shown to meet the penetration criterion by a validated analysis method that includes uncertainty factors on the container thickness or by test.

        (2) Composite containers are shown to meet this criterion by establishing that the composite container can sustain DUL (verified by analysis combined with coupon or hardware element test data) with the worst-case impact damage from the released part.

    b. Release or failure of the contained part because of a flaw does not result in a catastrophic hazard.

    c. The enclosure structure and supports meet the following:

        (1) Fracture control requirements listed in this NASA Technical Standard.

        (2) Perform their intended functions if impacted by the loose part, fragments, or contents of the part.

    d. Assessment of containers with mechanically secured closures shows the design is at least one fault tolerant against release of the contents.

Consider all sources of energy available to a contained part during a containment analysis.

If the part contains hazardous materials or fluids, to satisfy item 6.2.2.b (above), the containment assessment also establishes that no hazardous materials or part fragments are released that result in a catastrophic hazard. Also note that impact with a composite enclosure is to be considered during fracture control classification and assessment of the enclosure.

6.2.3 NFC Fail-Safe

Parts with sufficient structural redundancy that may fail because of the presence of a flaw may be designated NFC via the fail-safe category.

Satisfy all of the following items to classify a part as NFC fail-safe to meet requirement [FCR 8] section 6.2.c in this NASA Technical Standard:

a. Documented assessment establishes that loss of any load path does not result in a catastrophic hazard and that risk of loss of the structural redundancy because of multi-site fatigue or damage of redundant load path structures from any source during the service life of the structure is not a credible concern.

b. Failure of the part does not generate pieces or debris that would violate the NFC low-released mass requirements in section 6.2.1 in this NASA Technical Standard.

c. After the loss of any load path, there is sufficient remaining structural capability to safely sustain all resulting redistributed loads and environments (including dynamic response changes) until termination of the mission or until such time as the part is inspected and refurbished.

(1) For NFC composite or bonded parts that may be impacted by an NFC fail-safe part, establish that the impacted NFC composite or bonded parts can sustain DUL verified by analysis combined with coupon or hardware element test data, while being subjected to the worst-case impact damage from the NFC fail-safe part.

(2) For any remaining NFC composite or bonded structure of this fail-safe part, establish that the remaining structure can sustain DLL. This is verified by analysis combined with coupon or hardware element test data with the worst-case impact damage from the NFC fail-safe part.

(3) For fracture critical composite or bonded parts that may be impacted by an NFC fail-safe part, include the worst-case impact damage caused by the NFC fail-safe part in the DTA and RTD during evaluation of the fracture critical parts as described in section 7.4 in this NASA Technical Standard.

For metallic structures, verification by analysis is sufficient.

d. Re-flight hardware is verified by visual inspection or other means to be intact and free of structural anomalies before being re-flown.

The possible consequences of the release of redundant parts need to be assessed. Failure of a redundant part may create impact with an adjacent part and is to be considered during fracture control classification and assessment of the adjacent part.

6.2.4 NFC NHLBB Pressurized Components

This classification is intended for metallic pressure-bearing walls of containers, trapped volumes, lines, fittings, valves, regulators, filters, bellows, or other pressurized hardware that transfer non-hazardous fluid under pressure and that would leak down in the presence of a flaw rather than burst when used as intended. Typically, these parts are produced under process control in large quantities, are identical parts, and are subjected to NDE and qualification testing to ensure the parts are reliable and present a low risk of containing detectable flaws that result in crack growth related to environmental, loading, or other conditions. Also, this classification is intended for hardware designed to carry primarily pressure loads. This hardware is usually designed with appropriate supports, brackets, or relief loops such that the hardware is not subject to significant structural loads. In this classification, the leakage of the fluid is not allowed to create a catastrophic hazard. This section does not apply to the hardware types addressed in section 7.2 in this NASA Technical Standard.

Satisfy all of the following items to classify a part as an NFC NHLBB component to meet requirement [FCR 8] section 6.2.d in this NASA Technical Standard:

a. The pressurized item satisfies the LBB definition in this document at MDP.

b. The leak does not cause a catastrophic hazard nor release hazardous fluid.

c. As the hardware item leaks down, there is no repressurization or continued pressure cycles that could lead to continued fatigue or crack growth related to EAC or SLC.

d. The hardware is manufactured from metal alloys that are not susceptible to crack growth related to EAC or SLC in the applicable environment and that are typically used for pressurized systems, using processes that have been established by reliability or inspections of many similar parts to be extremely unlikely to produce parts with a flaw exceeding process specifications.

e. Associated structure supporting the pressurized hardware also meets fracture control requirements.

f. Hardware does not have an impervious barrier, coating, etc., on either the interior or exterior surfaces that inhibits leakage.

g. Re-flight hardware is inspected for leaks before repressurization and/or before being re-flown.

Note that leaking hardware may present unacceptable impacts on program mission success. Catastrophic hazards for LBB assessment include unacceptable dilution or toxicity of breathing environment, increases in oxygen or flammable material beyond flammability limits, or loss of a safety-critical function.

When LEFM is applicable, an acceptable approach to LBB for metallic alloys is to show by analysis that a worst-case surface crack will grow into a through-the-thickness crack without unstable crack propagation. This presumes the hardware manufacturing process has no credible risk of producing initial flaws longer than the crack, and leakage through the crack is shown to reduce pressure before loadings could grow the crack to cause fracture. The analysis, taking into account applied loads and residual stress effects, shows that the crack will leak and not be unstable. Additional guidance on analysis and leakage is available in API 579-1/ASME FFS-1, Fitness-for-Service.

6.2.5   NFC Low-Risk Parts

The low-risk classification is intended for parts that are extremely unlikely to contain or develop critical flaws because of (1) extremely low likelihood of flaws being induced by manufacturing processes, environmental effects, or service events and (2) large structural margins.

Satisfy all of the items in section 6.2.5.a for metallic parts or all of the items in section 6.2.5.b for composite or bonded hardware to classify a part as an NFC low-risk part to meet [FCR 8] section 6.2.e in this NASA Technical Standard:

    a.   Metallic parts are classified as low risk, provided the documented assessment shows they meet the following:

        (1) The part is manufactured from materials with well-characterized strength and ductility properties using processes that have been established by inspections to be extremely unlikely to produce parts with flaws and that have been shown not to fail because of brittle fracture.

        (2) Metallic parts have a material property ratio of $K_{Ic}/F_{ty} \geq 1.66 \sqrt{mm}$ ($0.33 \sqrt{in}$) and do not have sensitivity to EAC, SLC, or stress corrosion cracking as defined in NASA-STD-6016.

        (3) Aluminum parts are not loaded in the short transverse direction if this dimension (from the raw stock part) is greater than 7.62 cm (3 in).

        (4) Parts have total net-section stresses, e.g., maximum principal or von Mises, whichever is larger, at limit load that are less than 30 percent of the ultimate strength.

        (5) One of the following is satisfied:

A. Perform a fatigue analysis that results in a minimum service life factor of 4 with a factor of 1.5 on local cyclic stresses.

For metallic parts addressed in 6.2.5.a.5.A in this NASA Technical Standard, the part should meet conventional fatigue, accounting for notch and mean stresses, with 4 lifetimes and 1.5 on alternating stress.

B. Perform a damage tolerance analysis that results in a minimum of 4 complete service lives with a factor of 1.5 on alternating stress using a 0.127-mm (0.005-in) initial crack that conservatively accounts for the effects of notches and mean stress.

b. Composite or bonded hardware is classified as low risk, provided the documented assessment shows it meets the following, based on the flaws identified by the RTD performed in accordance with section 7.4.3 in this NASA Technical Standard:

(1) The part residual strength with the largest RTD flaw can sustain DUL verified by analysis combined with coupon or hardware element test data.

(2) The part limit strain with the RTD established flaw size is below the no-growth threshold strain established by test.

(3) Re-flight hardware is verified by visual inspection or other means to show the hardware is intact and free of structural anomalies before being re-flown.

Note that metallic welds and castings are manufacturing processes that may be likely to contain critical flaws, and therefore, they do not qualify as low-risk parts unless inspection data establish they have no flaws that can grow, i.e., the crack stress intensity factor is below threshold, including environment and residual stress effects.

For metallic parts addressed in item 6.2.5.a (above), the net-section stresses are to be computed based on strength-of-materials theory. An example of the net-section stress calculation for combined tension and bending stress is detailed in the NASGRO® User's Manual, Appendix B, in the beginning pages, except no crack or epsilon factor is used for this NFC low-risk application. For complex parts where finite element results are obtained that may include stress concentrations and stress gradients, the net-section stresses are to be computed by integrating the stress distribution and dividing by the area for the sectional area being assessed.

6.2.6   NFC Documented Non-Hazardous Failure Mode

Provide documentation establishing that a hazard assessment has been performed and that there are no credible catastrophic hazards resulting from failure of the part caused by a flaw to classify a part as NFC Documented Non-Hazardous Failure Mode to meet requirement [FCR 8] section 6.2.f in this NASA Technical Standard.

Note that this category is significantly different from the Exempt classification in section 5. Exempt parts are nonstructural and have no hazardous concerns or failure modes. This category may have structural or non-structural parts that are to be addressed by a documented hazard assessment that establishes no credible catastrophic hazards exist for the failure modes identified.

For composite or bonded parts classified as NFC Documented Non-Hazardous Failure Mode according to this section may not be required to meet all the requirements in section 6.3 in this NASA Technical Standard. Guidance from the RFCB may be necessary.

6.3 Additional Activities for Composite or Bonded NFC Hardware

Composite or bonded hardware classified as NFC require activities (detailed below) to be performed and then documented in the FCSR in addition to the other activities for the specific NFC category.

[FCR 9] NFC composite or bonded parts that satisfy requirements for classification in a specific category in sections 6.1 and 6.2 in this NASA Technical Standard shall also comply with all of the following items:

a. For parts classified as NFC low risk, develop the following:

(1) A DTA in accordance with section 7.4.1 in this NASA Technical Standard.
(2) An IDMP in accordance with section 7.4.2 in this NASA Technical Standard.
(3) An RTD in accordance with section 7.4.3 in this NASA Technical Standard.

b. For NFC parts not classified as low risk, perform the following:

(1) Define and quantify the flaws from any source that may occur to the hardware during its service life, considering all applicable flaw detection and mitigation strategies that are implemented for the flight hardware.

(2) Develop an IDMP in accordance with section 7.4.2 in this NASA Technical Standard.

c. Perform NDE after completion of all manufacturing processes (or after proof test, if a proof test is performed) in accordance with section 8.1.2 in this NASA Technical Standard, with the following clarifications:

(1) No NDE is required for NFC low-released mass parts.

(2) No NDE is required for NFC contained parts.

No NDE is required because there is no credible catastrophic hazard for these two specific categories.

d.   Meet the traceability requirement of section 8.2 in this NASA Technical Standard [FCR 21].

e.   Meet the material selection and usage requirement of section 8.3 in this NASA Technical Standard [FCR 22].

[Rationale: Parts classified as NFC that also contain composite or bonded materials need additional precautions to provide mitigation for undetected damage. These parts can be classified as NFC.]

Use of an alternative approach requires unique rationale and approval by the RFCB as described in section 10 [FCR 26] in this NASA Technical Standard.

The assessments in sections 7.4.1 and 7.4.2 in this NASA Technical Standard rely on NDE and material controls, including traceability requirements as prescribed in section 8 in this NASA Technical Standard to address hazards.

Traceability for NFC composite or bonded parts is somewhat unique relative to NFC metallic parts. While metallic parts usually have a specification for providing minimum properties throughout the part, composite or bonded parts are composed of elements that may have specifications, but the properties after combination of these elements are often unique to the hardware being produced.

NASA-STD-5001 requires a proof test for all composite or bonded structures.

7.   ASSESSMENT OF FRACTURE CRITICAL PARTS

7.1   Fracture Critical Parts

[FCR 10] Parts shall be classified as fracture critical unless one of the following is met:

a.   There is no credible possibility for a flaw in the part to cause failure during the lifetime of the part.

b.   Part failure does not result in a credible catastrophic hazard.

[Rationale: Parts that do not meet one of the above criteria require mitigation to preclude catastrophic failure. Classification as fracture critical denotes the need for knowledge of the sensitivity of the part to flaws or damage, an adequate screening of parts for flaws or damage and protection from damage, and traceability to assure a high-quality aerospace part is produced.]

Parts that are fracture critical require risk mitigation activities to provide assurance that flaw or damage sensitivity is understood relative to flaw screening or qualification and acceptance testing and material processing parameters.

The methods in this section are based on NASA's experience base, established approaches, industry standards, or aerospace standards. Any deviations or omissions of elements in the activities or approaches described in this section constitute an alternative approach that is to satisfy the requirements in section 10 [FCR 26] in this NASA Technical Standard.

In addition to assessments discussed in the subsequent subsections, fracture critical parts are subject to flaw screening, traceability, and material selection requirements in accordance with section 8 in this NASA Technical Standard. Documentation of the approaches to implementation and the results of implementation activities are discussed in section 9 of this NASA Technical Standard for the FCP and FCSR requirements.

    a. *Parts are fracture critical unless one of the following is met:*

        (1) The part is exempt in accordance with section 5 in this NASA Technical Standard.
        (2) The part is NFC in accordance with section 6 in this NASA Technical Standard.

    b. *Fracture critical parts are to comply with one of the following applicable items:*

        (1) Established approaches for the specific hardware types in accordance with section 7.2 in this NASA Technical Standard.

        (2) General approach for fracture critical metallic parts assessment in accordance with section 7.3 in this NASA Technical Standard.

        (3) General approach for fracture critical composite or bonded hardware assessment in accordance with section 7.4 in this NASA Technical Standard.

        (4) Optional approaches for fracture critical parts in accordance with section 7.5 in this NASA Technical Standard.

        (5) Satisfy requirements in section 10 in this NASA Technical Standard for an alternative approach.

    c. *Fracture critical parts are also to comply with the following items:*

        (1) Satisfy flaw screening, traceability, and material requirements in section 8 in this NASA Technical Standard.

        (2) Satisfy documentation requirements in section 9.1 in this NASA Technical Standard.

    d. *A part should always be classified as fracture critical if there is doubt or concern to establish that it is not fracture critical.*

    e. *Parts that are often determined to be fracture critical include but are not limited to: rotating hardware that does not satisfy this NASA Technical Standard's section 6.1.3 requirements, hazardous fluid containers, pressure systems that contain hazardous fluids (such as liquid rocket engine systems), and pressurized structures (such as propellant tank structures), primary thrust structure (unpressurized), solid rocket motor cases and nozzles, and habitable modules.*

    f. *Pressure vessels, as defined in section 3.2 of this NASA Technical Standard, are fracture critical.*

    g. *Fracture critical parts receive additional attention beyond the standard structural and quality assurance assessments normally given to spaceflight hardware. These additional activities include the following:*

        (1) *Either an approved set of prescribed activities deemed to be sufficient to mitigate the risk of failure because of a flaw (established approaches and optional approaches) or a damage tolerance assessment (analysis, test, or both) to show life requirements are met in the presence of flaws.*

        (2) *Screening of parts for flaws.*

        (3) *Traceability of the parts.*

        (4) *Material requirements.*

        (5) *Documentation of the assessment and hardware implementation process.*

7.2   Established Approaches for Specific Fracture Critical Hardware Types

[FCR 11] Each fracture critical part that is described by a specific hardware type in the following list shall comply with the established approach given in one of the following items:

    a. Fracture critical metallic pressure vessels comply with section 7.2.1 in this NASA Technical Standard.

    b. Fracture critical COPVs and composite overwrapped pressurized fluid containers comply with section 7.2.2 in this NASA Technical Standard.

    c. Other fracture critical pressure vessels and pressurized fluid containers comply with section 7.2.3 in this NASA Technical Standard.

d.   Fracture critical lines, fittings, and other pressurized components comply with section 7.2.4 in this NASA Technical Standard.

e.   Fracture critical habitable structures and volumes comply with section 7.2.5 in this NASA Technical Standard.

f.   Fracture critical pressurized structures comply with section 7.2.6 in this NASA Technical Standard.

g.   Fracture critical rotating hardware complies with section 7.2.7 in this NASA Technical Standard.

h.   Fracture critical fasteners comply with section 7.2.8 in this NASA Technical Standard.

i.   Fracture critical shatterable components and structures comply with section 7.2.9 in this NASA Technical Standard.

j.   Fracture critical tools, mechanisms, and tethers comply with section 7.2.10 in this NASA Technical Standard.

k.   Fracture critical batteries comply with section 7.2.11 in this NASA Technical Standard.

[Rationale: Parts that comply with this requirement have had sufficient activities performed to establish adequate risk mitigation of failure caused by the presence of a flaw or crack-like defect.]

There are currently no predefined approaches for pressure vessels or pressurized fluid containers that are qualified under a different code/standard than ANSI/AIAA S-080 or ANSI/AIAA S-081, such as the ASME Boiler and Pressure Vessel Code, Section VIII, Divisions 1 or 2, or the United States Department of Transportation (DOT) Code of Federal Regulations Title 49, Transportation. These codes/standards do not impose the structural integrity activities needed for damage tolerance that are specified in ANSI/AIAA S-080 and ANSI/AIAA S-081. The approaches used by these ASME, DOT, and other industrial codes/standards to certify vessels do not include damage tolerance. In addition, service fluid, temperature, mounting, vibration, or vacuum requirements consistent with aerospace environments are not addressed in these codes/standards. Damage tolerance is required for commercially available off-the-shelf (COTS) pressure vessels. Pressure vessels certified to ASME, DOT, and other industrial codes/standards with failure modes where leakage would not result in a catastrophic hazard (some examples of leakage resulting in catastrophic hazards are: toxic release, asphyxiation hazards, flammable mixture release, thrust loading on the pressure vessel mounting or surrounding structure that results in loss of structural margin or the need for operational modifications, or loss of critical system function) may be proposed for acceptance without damage tolerance assessment (in combination with other activities) by developing an alternative approach as required in FCR [26], section 10 of this NASA Technical Standard.

Equivalence means that damage tolerance life analysis or test requirements in sections 7.2.1 or 7.2.2 in this NASA Technical Standard are also applied in modified form for a vessel meeting section 7.2.3 in this NASA Technical Standard. Equivalence does not mean other types of assessment, such as fatigue calculations or cycle test, can be substituted for the damage tolerance methodology detailed in sections 7.2.1 and 7.2.2 in this NASA Technical Standard.

7.2.1   Fracture Critical Metallic Pressure Vessels

This category pertains to pressure vessels that are designed to meet ANSI/AIAA S-080-1998, Space Systems - Metallic Pressure Vessels, Pressurized Structures, and Pressure Components. Fracture critical metallic pressure vessels meeting other codes/standards are addressed in section 7.2.3 in this NASA Technical Standard.

Pressure vessels as defined by NASA are always fracture critical. For reference only, the definition of pressure vessels is repeated as guidance below.

> *Pressure Vessel: A container designed primarily for pressurized storage of gases or liquids and that also performs any of the following:*

- *Contains stored energy of 19,307 J (14,240 ft-lb) or greater based on adiabatic expansion of a perfect gas.*

- *Stores a gas that will experience an MDP greater than 690 kPa (100 psia).*

- *Contains a fluid (gas and/or liquid) in excess of 103 kPa (15 psia) that will create a hazard if released.*

Fracture critical metallic pressure vessels are to comply with ANSI/AIAA S-080-1998, with tailoring as specified below in items a through k to meet requirement [FCR 11] section 7.2.a in this NASA Technical Standard.

Subsequent versions of ANSI/AIAA S-080 with modifications that implement the technical content as mandated in this section may be used with the approval of the RFCB.

a.   Describe the damage tolerance assessment approach in the FCP in accordance with section 4.1 [FCR 1] in this NASA Technical Standard.

b.   All occurrences of the following terms in ANSI/AIAA S-080-1998 are replaced with the terms having meanings as specified below:

(1) All occurrences of "maximum expected operating pressure" and "MEOP" are substituted with "maximum design pressure" and "MDP" as terms in this NASA Technical Standard in section 3.2.

(2) The word "nominal" is replaced with the word "average" in all ANSI/AIAA S-080-1998 sections except 4.7.2.

(3) All occurrences of the term "service life" have the meaning defined in this NASA Technical Standard in section 3.2 for "service life."

c. The ANSI/AIAA S-080-1998 requirements in section 5.1, Approach A, Path 2, as detailed in section 5.1.2 of that document are followed for all the metallic pressure vessels addressed by that section with the modifications specified in this section of this NASA Technical Standard.

d. ANSI/AIAA S-080-1998 section 4.2.7 safe-life requirements are met with the following modifications:

(1) The safe-life assessment analysis and test assessments are to encompass and represent the worst-case flaw location, shape, aspect ratio, and orientation.

(2) The process for selecting the worst-case flaw location, shape, aspect ratio, and orientation is based on vessel stress/strain response, material strength, and crack growth properties and documented in the analysis report.

(3) The assessment determining the worst-case flaw location, shape, aspect ratio, and orientation includes all regions of the pressure vessel, including the boss and any internal and external attachments.

(4) The safe-life assessment analysis and test loading spectra are to include all loadings experienced during the service life, including those specified in this NASA Technical Standard in section 7.3.1, unless the RFCB approves the exclusion of specific loadings as insignificant for a component assessment.

For example, with approval of the RFCB, service life loadings that affect the safe-life of a particular region of the vessel by less than 5 percent may be excluded from the safe-life assessment of these regions.

(5) The assessments are to show that all safe-life requirements are met for the entire mission service life.

The mission service life includes all of the hardware activities included in the hardware mission as defined in NPR 7120.5, for the duration of the service life as defined in section 3.2 in this NASA Technical Standard. If the mission service life includes periodic "depot" intervals (opportunities for inspection) with fully qualified screening inspections that ensure that acceptable hardware has sufficient life, including the service life factor, to reach the next "depot" evaluation, this "depot" interval-based service-life approach may be proposed as an alternative approach by meeting the requirements in section 10 of this NASA Technical Standard.

e. If the AIAA S-080-1998 section 4.2.7 analysis option to show safe-life is planned, apply the following modifications to the requirements:

(1) Obtain pre-approval by the RFCB for all crack growth computer analysis programs other than NASGRO®.

(2) If the analysis ability to simulate crack growth is invalidated by plasticity or other effects, the assessment is performed by test.

(3) If NASGRO® is used, either set B_k to zero, or set B_k such that the stress intensity factor for the part thickness is less than or equal to the critical stress intensity value with approval of the Technical Authority or the RFCB.

(4) Establish that the assessed parts survive 4 lifetimes without failure (hazardous leak or fracture instability) by analyses that assess all applicable effects causing crack growth as a result of cyclic loadings.

    A. If the loading sequence of high/low loads is unknown, then damage tolerance analysis is to show that the stress intensity factor at limit load is less than the critical stress intensity factor or residual strength at the end of 4 lifetimes.

    B. If the service lifetime is a single event or the fatigue crack growth is small relative to the critical crack size (initial and critical cracks are of similar size), the analysis is to establish one of the following:

        i. Reserve capability against fracture by meeting either a lower bound critical stress intensity factor or residual strength at the end of 4 lifetimes.

        ii. A factor of 1.4 on critical stress intensity factor or residual strength after 1 lifetime.

        Assessments of metallic alloys that are susceptible to crack growth related to SLC or EAC during the service life are addressed in item (6) below.

(5) Use critical stress intensity factor and cyclic threshold stress intensity range (ΔK_{th}) values that are less than or equal to the average values.

(6) For metallic alloys susceptible to EAC or SLC or both, satisfy all of the following:

    A. Use the lower bound value of stress intensity factor threshold for assessment of EAC (K_{EAC} or K_{IEAC} as appropriate) and SLC if the material exhibits these behaviors in the application conditions.

APPROVED FOR PUBLIC RELEASE – DISTRIBUTION IS UNLIMITED

    B. Show that the applied stress intensity factor related to the largest service load is smaller than the lower bound stress intensity factor thresholds determined in item A above at the end of 4 lifetimes.

    f.   When performing proof testing in accordance with ANSI/AIAA S-080-1998 sections 4.2.7, 4.6.4, and/or 5.1.2.4, the duration of the proof test loading is minimized while also meeting the requirement to verify the pressure stability.

    g.   If the AIAA S-080-1998 section 4.2.7 testing option to show safe-life is planned, requirements are to include the following items:

    (1) The testing approach and rationale are subject to both of the following:

       A. RFCB approval before implementation.
       B. Documentation in the FCP.

    (2) The testing is to show that the hardware meets the damage tolerance lifetime and failure condition requirements in ANSI/AIAA S-080-1998 as modified in this NASA Technical Standard for initial flaws in the worst location, aspect ratio, and orientation in conditions that account for the service environments.

    (3) Testing reports showing that the testing objectives have been achieved are documented in accordance with section 9.1 in this NASA Technical Standard and cited in the FCSR.

    h.   The ANSI/AIAA S-080-1998 section 5.1.2.6 Special Provision is not allowed.

    Pressure vessels as defined by NASA are always fracture critical.

    i.   Vessels with crack-like flaws that are induced during the manufacturing process are not accepted as flight hardware unless a process for remediation repair has been established and the Technical Authority approves the part and process.

    Refer to section 8.1.5 of this NASA Technical Standard for further requirements and guidance.

    j.   The ANSI/AIAA S-080-1998 requirements are subject to the following:

    (1) Quality assurance in section 4.6 of that document is supplemented by requirements in section 8 (and its subsections) in this NASA Technical Standard.

    (2) If there is a conflict with ANSI/AIAA S-080-1998, the ANSI/AIAA S-080-1998 requirements for quality assurance in section 4.6 of that document are superseded

by requirements in section 8 (and its subsections) of this NASA Technical Standard.

k. The ANSI/AIAA S-080-1998 requirements for fracture critical part documentation and reporting are subject to the following:

(1) Supplemented by requirements in section 9 (and its subsections) of this NASA Technical Standard.

(2) If there is a conflict with ANSI/AIAA S-080-1998, the ANSI/AIAA S-080-1998 requirements for fracture critical part documentation and reporting of that document are superseded by requirements in section 9 (and its subsections) of this NASA Technical Standard.

(3) The ANSI/AIAA S-080-1998 section 4.2.5 required stress analysis report and the section 4.2.7 safe-life analysis report are provided as part of the FCSR documentation.

Note that ANSI/AIAA S-080-1998 also addresses other hardware types, but only the metallic pressure vessel requirements as tailored in this section are applicable for this NASA Technical Standard.

7.2.2 Fracture Critical COPVs and Composite Overwrapped Pressurized Fluid Containers

This category pertains to composite overwrapped pressure vessels and pressurized fluid containers that are designed to meet ANSI/AIAA S-081-2000, Space Systems - Composite Overwrapped Pressure Vessels (COPVs). Composite overwrapped pressurized fluid containers are pressurized parts with a composite structure fully or partially encapsulating a metallic liner and that do not meet the definition of a pressure vessel. Fracture critical COPVs and composite overwrapped pressurized fluid containers meeting other codes/standards are addressed in section 7.2.3 in this NASA Technical Standard.

For fracture critical COPVs and all other fracture critical composite overwrapped pressurized fluid containers with a metallic liner, show compliance with ANSI/AIAA S-081-2000, with modifications as specified below in items a through m, to satisfy requirement [FCR 11] section 7.2.b in this NASA Technical Standard.

Subsequent versions of ANSI/AIAA S-081 with modifications that implement the technical content as mandated in this section may be used with the approval of the RFCB.

a. The ANSI/AIAA S-081-2000 requirements are followed for the assessment and qualification of all the composite overwrapped vessels and composite overwrapped pressurized fluid containers with metallic liners addressed by this section, regardless of the vessel fluid pressure, energy, or hazardous nature, with the modifications specified in this section in this NASA Technical Standard.

b.   The damage tolerance assessment approach is described in the FCP in accordance with section 4.1 [FCR 1] in this NASA Technical Standard.

c.   All occurrences of the following terms in ANSI/AIAA S-081-2000 are replaced with the terms having meanings as specified below:

(1)   All occurrences of "maximum expected operating pressure" and "MEOP" are substituted with "maximum design pressure" and "MDP" as defined in this NASA Technical Standard in section 3.2.

(2)   The word "nominal" is replaced with the word "average" in all ANSI/AIAA S-081-2000 sections.

(3)   All occurrences of the term "service life" are to have the meaning defined in this NASA Technical Standard in section 3.2 for "service life."

d.   ANSI/AIAA S-081-2000 section 4.2 requirements are met with the following modifications:

(1)   In ANSI/AIAA S-081-2000 section 4.2, the damage tolerance, i.e., safe-life, approach (b) is the only acceptable approach.

(2)   The ANSI/AIAA S-081-2000 section 4.2.7 safe-life requirements are met with the following modifications:

A.   The safe-life assessment analysis and test assessments are to encompass and represent the worst-case flaw location, shape, aspect ratio, and orientation.

B.   The process for selecting the worst-case location, shape, aspect ratio, and orientation is based on liner stress/strain response and material strength and crack growth properties and documented in the analysis report.

C.   The assessment determining the worst-case location, shape, aspect ratio, and orientation includes all regions of the liner and boss, including the shear region of the boss and any internal and external attachments.

D.   The safe-life assessment analysis and test loading spectra are to include all loadings experienced during the service life, including those specified in this NASA Technical Standard in section 7.3.1, unless the RFCB approves the exclusion of specific loadings as insignificant for a component assessment.

For example, with approval of the RFCB, service life loadings that affect the safe-life of a particular region of the liner, boss, or shear region of the boss by less than 5 percent may be excluded from the safe-life assessment of these regions.

E. The assessments are to show all safe-life requirements are met for the entire mission service life.

The mission service life includes all of the hardware activities included in the hardware mission as defined in NPR 7120.5, for the duration of the service life as defined in section 3.2 in this NASA Technical Standard. If the mission service life includes periodic "depot" intervals (opportunity for inspection) with fully qualified screening inspections that ensure acceptable hardware has sufficient life, including the service life factor, to reach the next "depot" evaluation, this "depot" interval-based service-life approach may be proposed as an alternative approach by meeting the requirements in section 10 in this NASA Technical Standard.

(3) Autofrettage is included in the service life unless liner NDE is performed after autofrettage.

(4) The assessment of crack growth related to the autofrettage cycle is determined by test in accordance with ANSI/AIAA S-081-2000 section 5.2.1, unless prior approval is provided by the RFCB for an analytical approach.

e. When performing analysis to show safe life for linearly responding portions of the metal liner in accordance with ANSI/AIAA S-081-2000 section 4.2.7, apply the following modifications:

(1) Obtain pre-approval from the RFCB for all crack growth computer analysis programs other than NASGRO®.

(2) If the analysis ability to simulate crack growth is invalidated by plasticity or other effects, the assessment is performed by test.

(3) If NASGRO® is used, either set B_k to zero, or set B_k such that the stress intensity factor for the part thickness is less than or equal to the critical stress intensity value with approval of the Technical Authority or the RFCB.

(4) The analysis shows that the parts survive 4 service lives without failure by assessments that address all applicable effects causing crack growth as a result of cyclic loading, using the following criteria:

Assessments of metallic alloys that are susceptible to crack growth related to SLC or EAC during the service life are addressed in item 6 below.

A. If the loading sequence of high/low loads is unknown, then damage tolerance analysis is to show that the stress intensity factor at limit load is less than the critical stress intensity factor or that the part applied load does not exceed the residual strength at the end of 4 lifetimes.

B. If the service lifetime is a single event or the amount of fatigue crack growth is small relative to the critical crack size for unstable crack growth, the analysis is to show reserve capability against fracture by meeting either of the following:

i. A lower bound critical stress intensity factor or residual strength at the end of 4 lifetimes.

ii. A factor of 1.4 on critical stress intensity factor or residual strength after 1 lifetime.

(5) Use critical stress intensity factor and cyclic threshold stress intensity range (ΔK_{th}) values that are less than or equal to the average values.

(6) For metallic alloys susceptible to EAC or SLC or both, satisfy all of the following:

A. Use the lower bound value of stress intensity factor threshold for assessment of EAC (K_{EAC} or K_{IEAC} as appropriate), and SLC if the material exhibits these behaviors in the application conditions.

B. Show that the applied stress intensity factor related to the largest service load is smaller than the lower bound stress intensity factor thresholds determined in item A above at the end of 4 lifetimes.

f. When performing proof testing in accordance with ANSI/AIAA S-081-2000 section 5.1.2, the duration of the proof test loading is minimized while also meeting the requirement to verify the pressure stability.

g. When performing assessment to show safe-life by test for non-linear response of the metal liner in accordance with ANSI/AIAA S-081-2000 section 5.2.1, apply the following items:

(1) The testing approach and rationale are subject to both of the following:

A. Provided to the RFCB for approval before implementation.
B. Documented in the FCP.

(2) The testing is to show that the hardware meets the damage tolerance lifetime and failure condition requirements in ANSI/AIAA S-081-2000 as modified in this

NASA Technical Standard for initial flaws in the worst location and orientation in conditions that account for the service environments.

(3) Testing reports showing that the testing objectives have been achieved and are documented in accordance with section 9.1 of this NASA Technical Standard and cited in the FCSR.

h. ANSI/AIAA S-081-2000 section 4.2.10 damage control requirements are to include the section 4.2.10.2.1 protective cover approach with the following additional requirements:

(1) The covers are required regardless of the COPV burst factor, wall thickness, hazardous or nonhazardous nature of the fluid, or energy content.

(2) If the vessel is exposed to risk of damage during any parts of the service life where the initially applied covers are not present, additional damage controls are selected from the options in ANSI/AIAA S-081-2000 section 4.2.10.

The additional damage controls referenced in item h.2 above may be needed if protective covers are removed before launch and the vessel has risk of damage during the remainder of its service life or if there are different risks to the vessel during its service life because of environments or service loadings that also need to be addressed by damage controls. All options, separately or in combination, may be used, including specialized covers for flight conditions.

i. Apply the following items if the composite overwrap is constrained by external structure or if it is part of a load path supporting the COPV for service life loads other than pressure loads:

(1) Perform an assessment validated by testing that shows the overwrap with the external structure loads meets all strength, fatigue, and life requirements in ANSI/AIAA S-081-2000.

(2) The assessment is to include effects of damage conditions that are not screened by the protections imposed in accordance with ANSI/AIAA S-081-2000 section 4.2.10 with the modifications in this section of this NASA Technical Standard.

j. Vessels with crack-like flaws in the metal liner that are induced during the manufacturing process are not accepted as flight hardware unless a process for remediation repair has been established and the Technical Authority approves the part and process.

Refer to section 8.1.5 in this NASA Technical Standard for further requirements and guidance.

k. Damage in other regions of the vessel may be repaired with an established, proven process if approved by the Technical Authority.

Refer to section 8.1.5 of this NASA Technical Standard for further requirements and guidance. This pertains to the repair of small manufacturing or cosmetic defects in the composite. There are no acceptable established processes for repairing impact damage to the composite overwrap. Accidental impacts that do not leave obvious visible damage indications are to be logged, the impact site assessed by qualified inspectors, and the hardware approved for use by the Technical Authority.

  l. The ANSI/AIAA S-081-2000 requirements for quality assurance in section 4.5 in that document are supplemented and superseded by requirements in section 8 (and its subsections) in this NASA Technical Standard.

  m. The ANSI/AIAA S-081-2000 requirements for fracture critical part documentation and reporting are subject to the following:

    (1) Requirements in section 9 (and its subsections) of this NASA Technical Standard.

    (2) If there is a conflict with ANSI/AIAA S-081-2000, the requirements are superseded by requirements in section 9 (and its subsections) of this NASA Technical Standard.

    (3) The ANSI/AIAA S-081-2000 section 4.2.7 safe-life and analysis reports and the 4.2.10 Mechanical Damage Control Plan (MDCP) are provided as part of the FCSR documentation.

The entity responsible for delivery of the MDCP (NASA, prime contractor, or other subcontractors) determines who develops the MDCP, which is subject to RFCB approval.

7.2.3 Other Fracture Critical Pressure Vessels and Pressurized Fluid Containers

Satisfy the following for all other fracture critical pressure vessels and pressurized fluid containers that are not addressed in either section 7.2.1 or 7.2.2 in this NASA Technical Standard to satisfy requirement [FCR 11] 7.2.c in this NASA Technical Standard:

  a. Document the proposed approach in the FCP in accordance with section 4.1 [FCR 1] in this NASA Technical Standard and include the following:

    (1) A rationale for using a metallic pressure vessel, COPV, or composite overwrapped pressurized fluid container instead of one of the following:

    A. An all-metal pressure vessel that meets the requirements of section 7.2.1 in this NASA Technical Standard, or

    B. A COPV or composite overwrapped pressurized fluid container that meets the requirements of section 7.2.2 in this NASA Technical Standard.

(2) Describe the proposed approach that satisfies applicable requirements in items b, c, or d below in this section and the requirements in sections 8 and 9 in this NASA Technical Standard.

(3) Receive RFCB approval before implementing the proposed approach.

A rationale is required because detailed requirements for the approach have to be developed and documented in the FCP that satisfy the applicable requirements in b or c below and the guidance in this section, and the RFCB has to review and approval the proposed detailed approach. This presents a significant effort for the developer of the FCP and for the RFCB reviews.

b. The development approach is satisfied by comparison to requirements in sections 7.2.1 in this NASA Technical Standard for metallic pressure vessels or 7.2.2 in this NASA Technical Standard for COPVs and composite overwrapped pressurized fluid containers. The approach is to be equivalent to or an extension of all the requirements, including establishing that damage tolerance life is achieved without failure or leakage of the fluid, and provides equivalent risk mitigation of a catastrophic failure caused by flaws.

c. The proposed FCP approach for damage tolerance assessment of a fracture critical pressure vessel or pressurized fluid container that is all composite or has a non-metallic, i.e., an elastomeric, liner or other non-metallic components is to meet the general approach for fracture critical composite hardware in section 7.4 in this NASA Technical Standard and show that the damage tolerance required life is achieved without failure or leakage of the fluid.

d. The proposed FCP approach for damage tolerance assessment of a fracture critical all-metal pressurized fluid container is to meet the general approach for fracture critical metallic hardware in section 7.3 in this NASA Technical Standard and show that the damage tolerance life is achieved without failure or leakage of the fluid.

Note that if a fracture critical metallic "pressurized fluid container" is planned with attributes close to the definition of a pressure vessel, it may be advantageous to push it into the pressure vessel category to minimize later impacts as the project matures in case the initial design attributes increase.

There are currently no predefined approaches for pressure vessels or pressurized fluid containers that are qualified under a different code/standard than ANSI/AIAA S-080 or ANSI/AIAA S-081, such as the ASME Boiler and Pressure Vessel Code, Section VIII, Divisions 1 or 2, or the United States Department of Transportation Code of Federal Regulations Title 49, Transportation. These codes/standards do not impose the structural integrity activities needed for damage tolerance that are specified in ANSI/AIAA S-080 and ANSI/AIAA S-081. As a result, the approaches used by these codes/standards to certify vessels do not facilitate meeting damage tolerance requirements as required in this NASA Technical Standard. Equivalence means that damage tolerance life analysis or test requirements in sections 7.2.1 or 7.2.2 in the NASA Technical Standard are also applied in modified form for a vessel meeting section 7.2.3 in this NASA Technical Standard. Equivalence does not mean other types of assessment, such as fatigue calculations or cycle test, can be substituted for the damage tolerance methodology detailed in sections 7.2.1 and 7.2.2 in this NASA Technical Standard. Use of these codes/standards in combination with other activities may be proposed, however, as an alternative approach as described in [FCR 26] in section 10 in this NASA Technical Standard.

In addition, other pressure vessels and pressurized fluid containers may be developed that are not addressed by existing codes or standards. Examples may include composite pressure vessels/containers without a metal liner or rubber-lined composite pressure vessels/containers.

For these other fracture critical vessels/containers, a unique approach is developed and proposed in the FCP that establishes equivalent methods of addressing material, structural, qualification, acceptance, and related aspects such as those in the ANSI/AIAA S-080 or ANSI/AIAA S-081 standards to support the damage tolerance assessment. Equivalence with the AIAA pressure vessel standards may include assessments and testing that include materials aspects, loadings, stress analysis, strength, environment effects, stiffness, thermal response, life, quality assurance, repairs, NDE requirements, acceptance processes including proof and leakage testing, damage tolerance control plans, and damage tolerance assessments by analysis and/or testing, and documentation. However, use of analytical techniques to establish damage tolerance is generally considered insufficiently developed for composite pressure vessels. For all-composite pressure vessels, the approaches described for fracture critical composite hardware in section 7.4 in the NASA Technical Standard should be incorporated, in addition to applicable equivalent requirements in ANSI/AIAA S-081.

For other types of vessels/containers, it should also be noted that, in addition to the section 9 documentation in this NASA Technical Standard showing the approach proposed in the FCP has been met, section 9.1.3.1 in this NASA Technical Standard requires providing supporting detailed technical information to the RFCB upon request, including drawings, material and processing data, detailed stress analysis, and damage tolerance analyses that are needed to support the damage tolerance assessment.

Early involvement with the RFCB is suggested for any vessels/containers to be assessed by this section.

7.2.4 Fracture Critical Lines, Fittings, and Other Pressurized Components

For metallic fracture critical lines, fittings, and other pressurized components (hardware items that are part of a pressurized system, including valves, filters, regulators, heat pipes, and heat exchangers) that transfer hazardous fluids or when loss of pressurization results in a catastrophic hazard, to satisfy requirement [FCR 11] section 7.2.d in this NASA Technical Standard, meet either 7.2.4.a or 7.2.4.b (below).

a. Apply the following items (1) through (6) to parts where the only load of significance is related to pressure:

(1) The metallic material is not susceptible to crack extension related to EAC or SLC.

(2) Perform 100 percent inspection of all fusion joints in fracture critical pressure components using a qualified NDE method after proof test to inspect for the presence of unacceptable lack of penetration or other unacceptable conditions both on the surface and within the fusion joint.

(3) Reject any type of flaw indication in the final product that does not meet specification requirements.

NDE rejection indicates the need for formal review and part disposition.

(4) Proof test lines, fittings, joints, and other pressurized components or parts to a minimum of 1.5 times the MDP during individual acceptance or at the system level.

(5) An ECF less than 1.0 is not allowed without prior approval by the RFCB.

(6) Obtain RFCB approval that the part is manufactured using processes that have been established by reliability or by inspections of many similar parts to be extremely unlikely to produce parts with a flaw exceeding process specifications.

For loading (stresses) to be considered pressure dominant, all other loads (stresses) should be no greater than 20 percent of the pressure loads (stresses).

b. Satisfy section 7.3 in this NASA Technical Standard for parts that do not meet the criterion in 7.2.4.a.

Item 7.2.4.a is intended for hardware designed to carry primarily pressure loads. This hardware is designed with appropriate supports, brackets, or relief loops such that they are not subject to significant structural loads. Typically, these parts are produced under process control in large quantities, are identical parts, and are subjected to NDE and qualification testing to ensure the parts are reliable and present a low risk of containing detectable flaws that result in crack growth.

Pressurized components may have high pressures and energies, but this type of hardware is subject to high factors of safety imposed by other standards such as NASA-STD-5001. NASA-STD-5001 also requires implementation of AIAA S-080, which has a leak test requirement.

7.2.5 Fracture Critical Habitable Modules and Volumes

Satisfy the following for fracture critical habitable modules and volumes to meet requirement [FCR 11] section7.2.e in this NASA Technical Standard:

a. Establish that pressure shells are damage tolerant by satisfying sections 7.3 or 7.4 in this NASA Technical Standard for the appropriate material type.

b. Proof test pressure shells.

c. Perform post-proof test NDE of pressure shell welds.

d. Monitor and document operation to ensure that certification is not invalidated.

Proof tests are usually performed in the operational environment, or the test levels are adjusted via an ECF.

Proof test levels (factors) are defined either by structural requirements or those developed to provide flaw screening (section 8.1.3 [FCR 18] in this NASA Technical Standard). Section 8.2 in this NASA Technical Standard requires load history traceability for all fracture critical parts.

Flaw screening for the entire fracture critical structure is required in accordance with section 8 in this NASA Technical Standard. Pre-proof NDE is highly recommended to protect high-value structures and facilities.

A damage tolerance assessment considers the worst-case allowed weld joint peaking and mismatch effects (metallic structures) and residual stress effects (either by analysis or included as a part of material test data) for habitable structures and enclosures.

7.2.6 Fracture Critical Pressurized Structures

This section is intended for pressurized structures such as launch vehicle main propellant tanks that carry internal pressure and vehicle structural loads.

Satisfy the following for fracture critical pressurized structures to meet requirement [FCR 11] section 7.2.f in this NASA Technical Standard:

a. Proof test all flight articles.

b. For metallic pressurized structures, establish damage tolerance by satisfying section 7.3 in this NASA Technical Standard.

c. For metallic pressurized structures, perform post-proof test NDE in accordance with section 8.1.1 in this NASA Technical Standard, in addition to other necessary flaw screening required in section 8 in this NASA Technical Standard, in the following manner:

Standard NDE is acceptable.

(1) Welded regions where proof testing adequately screens for flaws are subject to the following:

A. Perform post-proof NDE (surface and volumetric) of all welded regions for the first flight article (as a minimum).

B. Also perform post-proof NDE of all affected weld regions (including those that are adequately screened for flaws by proof test) subjected to significant process, material, or vendor changes for the first flight article incorporating the significant changes.

(2) For welded regions where proof testing does not adequately screen for flaws, perform post-proof NDE (surface and volumetric) of all welded regions for all flight articles.

(3) All weld intersections, weld repair regions, and weld transition regions, including friction plug pull weld regions, are to receive post-proof NDE (surface and volumetric) for all flight articles.

d.   For composite or bonded pressurized structures, provide the damage tolerance approach and rationale to the RFCB for approval before implementation.

For composite or bonded pressurized structures, the requirements in section 7.4 in this NASA Technical Standard are a good starting point as a fracture control approach but will need enhancement to provide adequate protection against catastrophic hazard.

e.   For composite or bonded pressurized structures, perform post-proof NDE as described in section 8.1.2 in this NASA Technical Standard.

Proof tests are usually performed in the operational environment, or the test levels are adjusted via an ECF.

Proof test levels (factors) are defined either by structural requirements or by those developed to provide flaw screening (section 8.1.3 [FCR 18] in this NASA Technical Standard). Section 8.2 in this NASA Technical Standard requires load history traceability for all fracture critical parts. The use of pressurized structures should be monitored with documentation of the operational history to ensure that certification is not invalidated.

The proof test factor for these structures is a minimum of 1.05 in accordance with NASA-STD-5001. This may result in a high stress during proof and possible growth of large flaws in the structure during the proof test. In accordance with the guidance in section 8.1.3 in this NASA Technical Standard, the flaw size used in the life assessment of these structures in regions where the proof test is used for flaw screening needs to adequately account for possible flaw growth during the proof test (typically established by laboratory damage tolerance tests).

Although it may be difficult to obtain adequate flaw screening for all welded regions via a proof pressure test because of external vehicle loads, the proof test is designed to provide as much flaw screening for welds as is practical.

Flaw screening for the entire fracture critical structure is required in accordance with section 8 in this NASA Technical Standard. Pre-proof NDE is highly recommended to protect high-value structures and facilities.

Damage tolerant assessment considers the worst-case allowed weld joint peaking and mismatch effects (metallic structures) and residual stress effects (either by analysis or included as a part of material test data) for pressurized structures.

7.2.7 Fracture Critical Rotating Hardware

Satisfy the following for fracture critical rotating hardware, including rotating hardware that does not satisfy the conditions in NFC rotating hardware section 6.1.3 in this NASA Technical Standard, to meet requirement [FCR 11] in section 7.2.g in this NASA Technical Standard:

a. The rotating hardware is to satisfy the appropriate section 7.3 or section 7.4 in this Standard for the material type.

b. The rotating hardware is proofed by a spin test to a minimum rotational energy factor of 1.05, i.e., rotational test speed = $\sqrt{1.05\,\omega^2}$, and one of the following performed:

(1) Perform NDE in accordance with section 8.1 in this NASA Technical Standard before and after the spin proof test.

(2) Establish that the spin proof test adequately screens for flaws (section 8.1 in this NASA Technical Standard) and that this approach for flaw screening is approved by the RFCB.

Proof tests are usually performed in the operational environment, or the test levels are adjusted via an ECF.

7.2.8 Fracture Critical Fasteners

Satisfy the following for fracture critical fasteners to meet requirement [FCR 11] section 7.2.h in this NASA Technical Standard:

a. Design, fabricate, purchase, and implement fracture critical fasteners with all of the following attributes.

(1) Fasteners are fabricated from a metal with high resistance to stress corrosion cracking, as defined in MSFC-STD-3029.

(2) Fasteners are fabricated, procured, and inspected in accordance with NASA-STD-6008, and an equivalent military standard, NAS, proprietary, or commercial aerospace specification approved by the RFCB.

(3) The fastened joint complies with NASA-STD-5020 without joint separation in the nominal configuration.

(4) Fasteners have rolled threads and are assessed to demonstrate they meet the fatigue requirements in NASA-STD-5001.

(5) Fasteners manufactured from titanium alloys require additional coordination with the RFCB for approval.

Titanium alloys, such as Ti-6Al-4V (including annealed and STA conditions), cp-Ti, and other titanium alloys, have potential generic EAC or SLC failure modes that are to be addressed in the assessment with test data from flawed fasteners in the applicable service life environments.

(6) The fasteners are not made from a low fracture toughness alloy, as defined in section 3.2 in this NASA Technical Standard.

(7) Fasteners are not reworked or custom made unless the application is approved by the RFCB.

b. Include preload and its effect on flaws and cyclic stresses in the damage tolerance assessment.

c. Inspect all fracture critical fasteners by the eddy current NDE technique or use proof testing to screen for flaws.

d. Assume a flaw in the most critical location of a size consistent with NDE sensitivity or proof-test level in the damage tolerance analysis.

General NDE flaw sizes are given in NASA-STD-5009, Nondestructive Evaluation Requirements for Fracture Critical Metallic Components; but for specific guidelines on eddy current methodology, PRC-6509, Process Specification for Eddy Current Inspection, can be used as a reference.

e. Proof-load test inserts used in conjunction with fracture critical fasteners to a minimum factor of 1.2 after installation.

This would include, for example, inserts bonded or potted into composite and sandwich structures, as well as inserts installed into metallic structures. Note that composite structures require additional considerations, as given in section 7.4 in this NASA Technical Standard.

f. Store and control fracture critical fasteners after inspection or testing to keep them isolated from other fasteners.

7.2.9 Fracture Critical Shatterable Components and Structures

Satisfy the following for fracture critical shatterable components and structures to meet requirement [FCR 11] section 7.2.i in this NASA Technical Standard:

a. Follow the requirements contained in NASA-STD-5018 for fracture critical shatterable components in internal volumes.

b. Coordinate with the RFCB for fracture critical external shatterable components and structures.

7.2.10 Fracture Critical Tools, Mechanisms, and Tethers

The following are to be applied to fracture critical tools or mechanisms that are the only (no backup) means for performing a function where failure to perform the function would result in a catastrophic hazard or a tool or mechanism whose failure during use would, in itself, result in a catastrophic hazard. This classification includes safety-critical tethers.

Satisfy the following for fracture critical tools, mechanisms, and tethers to meet requirement [FCR 11] section 7.2.j in this NASA Technical Standard:

a. Perform NDE and damage tolerance assessment (as described in section 7.3 or section 7.4 in this NASA Technical Standard) for each fracture critical tool or mechanism to assure that flaws that could cause failure during use are not present.

b. Fracture critical springs require RFCB approval.

c. Qualification, design life verification, and acceptance testing are to comply with NASA-STD-5017, Design and Development Requirements for Mechanisms, for fracture critical mechanisms.

When NDE methods are not sufficient to screen for critical defects, rationale should be presented to the RFCB for approval that could include proof testing, statistical life testing, and other mechanical testing and analysis to provide further understanding of defect sensitivity in the part.

Springs should be designed to be fail-safe or redundant.

Tethers should be proof tested, inspected, and assessed for damage in accordance with applicable operational requirements.

Proof tests are usually performed in the operational environment, or the test levels are adjusted via an ECF. Other requirements such as NASA-STD-5001 provide proof test levels.

7.2.11 Fracture Critical Batteries

Satisfy the following for fracture critical batteries to meet requirement [FCR 11] section 7.2.k in this NASA Technical Standard:

a. Comply with JSC 20793, Crewed Space Vehicle Battery Safety Requirements.

b. Comply with section 7.5.5 in this NASA Technical Standard for fracture critical batteries.

7.3 General Approach for Fracture Critical Metallic Parts Assessment

[FCR 12] Each fracture critical metallic part that is not of a specific hardware type as described in section 7.2 in this NASA Technical Standard and is not approved by the RFCB as appropriate for an optional approach as described in section 7.5 in this NASA Technical Standard shall comply with one of the following item combinations: a and b; a and c; or a, b, and c:

a. Develop loading spectra by complying with section 7.3.1 in this NASA Technical Standard.

b. Perform assessment by analysis to comply with section 7.3.2 in this NASA Technical Standard.

c. Perform assessment by test to comply with section 7.3.3 in this NASA Technical Standard.

[Rationale: Fracture critical parts need activities performed to understand the sensitivity of the part if a flaw is present. These activities can range from a direct assessment of the part's capability with a flaw to acceptance tests that establish the part has sufficient capability to a combination of activities that provide sufficient information to mitigate the risk of failure related to undiscovered flaws.]

Use of an alternative approach requires unique rationale and approval by the RFCB as described in section 10 [FCR 26] in this NASA Technical Standard. The approaches in this requirement are the preferred approaches if followed completely.

Damage tolerant assessment used as the basis for acceptance of a fracture critical metallic part establishes all of the following:

- *The relevant critical failure mode for the part is identified.*

- *The appropriate load spectra are applied.*

- *The appropriate initial flaw size in a worst-case orientation based on the screening method implemented, in the worst location, is used.*

- *Conservative material data and analysis methods are used.*

- *One of the following (each of which is detailed in this section) is established:*

  - *The part has a minimum service life factor of 4.*

  - *The part is single loading event hardware and has a factor of 1.4 on critical stress intensity factor or residual strength.*

A damage tolerance assessment is performed to understand the sensitivity of a part to flaws. The requirement is necessary to mitigate risk of failure because of flaws that may still exist after implementation of flaw screening strategies. Fatigue-crack-growth empirical data have inherent scatter. When performing damage tolerance assessments, mean values are used, not a statistical lower bound. In addition, the prediction procedures have uncertainties related to the local stress levels, stress-intensity factor calculations, load spectra, and environmental effects. Errors in local stresses and stress-intensity factor calculations are grossly magnified when crack growth rates are evaluated while using the Paris growth law. Slight misjudgments of the spectrum can lead to large effects on crack growth. To account for all of these effects, a safety factor is applied on the predicted life. Thus, the life factor of 4 provides margin on uncertainties in analysis, prediction methodologies, and material property variations. The single load event factor of 1.4 on critical stress intensity factor, fracture toughness, or residual strength provides ultimate load capability with flaws that may go undetected and is representative of the requirements in NASA-STD-5001.

7.3.1 Loading Spectra

A loading spectrum is necessary for the damage tolerance life analysis or damage tolerance life test.

Develop loading spectra according to the following to satisfy requirement [FCR 12] section 7.3.a in this NASA Technical Standard:

    a. Include all anticipated significant loadings, both cyclic and sustained, for each fracture critical part throughout its service life.

    b. Include all load levels and the number of cycles or duration during the service life of the hardware, including proof test loads.

    c. Include the effects of the appropriate environment for each fracture critical part throughout its service life.

    d. Include the effects of preloads.

e. Include residual stresses and any weld joint discontinuities, such as peaking and mismatch, for cyclic and sustained loads during the service life of the hardware.

f. Include the influence of all coatings and barriers on pressure-loaded parts for any scenarios where pressure is assumed to decrease because of leakage from a crack.

g. Include the effects of impact loads and damage from mission environments, including but not limited to credible impacts from vehicle loss of external surface mass, MMOD, EVA inadvertent contacts, and EVA tool impacts during assessments of external structures and components.

Include the worst-case allowed or weld joint peaking and mismatch effects for damage tolerance assessments by analysis or test. The assessment analysis or test is to capture the effect of peaking and mismatch on stress gradients affecting crack growth and fracture. Standard tensile strength tests of ductile materials are not adequate to assess these conditions.

Proof load factors are listed in NASA-STD-5001 and may exist in program-specific requirements. Proof tests are usually performed in the operational environment, or the test levels are adjusted via an ECF.

7.3.2 Assessment by Analysis

Satisfy the following to perform assessment by analysis to meet requirement [FCR 12] section 7.3.b in this NASA Technical Standard:

a. Assume that the initial flaw that could be present and undetected in the part is the size and shape that is not screened by NDE, proof test, or process control and is in the worst location and orientation.

b. Use analysis methods and computer programs that are approved by the RFCB, e.g., NASGRO®, for predicting flaw growth, life, and critical flaw sizes.

Note that when the available analysis ability to simulate crack growth is invalid, assessment by test (section 7.3.3 in this NASA Technical Standard) is required.

c. Establish that the assessed parts survive 4 lifetimes without failure (hazardous leak or fracture instability) by analyses that assess all applicable effects causing crack growth as a result of cyclic loadings.

> (1) If the loading sequence of high/low loads is unknown, then damage tolerance analysis is to show that the stress intensity at limit load is less than the critical stress intensity factor or residual strength at the end of 4 lifetimes.

(2) If the service lifetime is a single event or the fatigue crack growth is small relative to the critical crack size (initial and critical cracks are of similar size), the analysis is to establish one of the following:

A. Reserve capability against fracture by meeting either a lower bound critical stress intensity factor or residual strength at the end of 4 lifetimes.

B. A factor of 1.4 on critical stress intensity factor or residual strength after 1 lifetime.

Assessments of metallic alloys that are susceptible to crack growth because of SLC or EAC during the service life are addressed in item 7.3.2.f below.

d. Use flaw growth rates that are greater than or equal to the average values without implementing retardation effects on flaw growth rates in the damage tolerance analysis.

e. Use critical stress intensity factor and cyclic threshold stress intensity range (ΔK_{th}) values that are less than or equal to the average values.

f. For metallic alloys susceptible to EAC or SLC or both, satisfy all of the following:

(1) Use the lower bound value of stress intensity factor threshold for assessment of EAC (K_{EAC} or K_{IEAC} as appropriate) and SLC if the material exhibits these behaviors in the application conditions.

(2) Show that the applied stress intensity factor related to the largest service load is smaller than the lower bound stress intensity factor thresholds determined in item (1) above at the end of 4 lifetimes.

Requirement 7.3.2.f is intended to preclude susceptible metallic alloy flight hardware from experiencing time-dependent, i.e. da/dt, crack growth.

g. If NASGRO® is used:

(1) B_k is either set to zero, or B_k is set such that K_c at the part thickness is less than or equal to the K_{Ic} value.

(2) Values of B_k resulting in $K_c > K_{Ic}$ require further understanding of the constraint condition for the crack situation and may be used with approval of the Technical Authority or RFCB.

h. Use fracture properties subject to all of the following:

(1) From sources or testing that are approved by the RFCB.

(2) Representative of the material process condition.

(3) Representative of weakest material orientation in the part (unless material orientation is fully traceable throughout the design and service life).

i. If material data needed for the damage tolerance assessment are not available, one of the following is to be accomplished:

(1) Obtain the data by material testing.

(2) If the source of the data to be used is from the literature, conduct an assessment to show that conservative results are obtained using that available data.

Section 8.1 in this NASA Technical Standard specifies flaw screening methods. The damage tolerance assessment is to address flaws that are not screened by the screening method applied to the flight hardware.

The NASGRO® computer program is an approved analysis tool for the damage tolerance life assessment of metallic spaceflight hardware. Other computer programs or analysis methods are acceptable with prior approval by the RFCB. The NASGRO® material database contains fracture mechanics properties for several materials that can be used with concurrence from the RFCB.

Standard NASA damage tolerance analyses are deterministic, and experience has shown these deterministic methods to be adequate. The probabilistic method uses knowledge (or assumptions) of the statistical variability of the damage tolerance variables to select criteria for achieving an overall success confidence level. Any proposed use of probabilistic damage tolerance analysis or criteria to meet fracture control requirements is considered an alternative approach as described in section 10 in this NASA Technical Standard and is approved by the RFCB on a case-by-case basis.

7.3.3 Assessment by Test

Perform assessment by test according to the following to satisfy requirement [FCR 12] section 7.3.c in this NASA Technical Standard:

a. Provide the approach and rationale to the RFCB for approval before implementation.

b. Document the approved approach in the FCP.

c. Perform the test(s) with initial flaws in the worst location and orientation.

d. Establish by testing that the components survive 4 lifetimes, including section 7.3.2.c.(1) and 7.3.2.c.(2) requirements in this NASA Technical Standard, without failure (leak or fracture instability).

Testing may be supplemented by analyses that, in conjunction or augmented by test correction factors, assess all applicable effects causing increased crack growth.

e. Test in conditions that account for the service environments.

f. A sufficient number of tests is performed to establish a representative result considering variability of material damage tolerance data.

The approved approach is to be documented in the FCP. Formal documentation in the FCP facilitates in-depth technical review and approval. Testing of coupons and pre-flawed structural elements representative of the flight hardware damage tolerance condition may be an acceptable approach to establish damage tolerance for metallic fracture critical parts. Together, the testing and any supplemental analyses are to establish that equivalent section 7.3.2 requirements in this NASA Technical Standard are met.

7.4 General Approach for Fracture Critical Composite or Bonded Hardware Assessment

[FCR 13] Each fracture critical composite or bonded part that is not of a specific hardware type as described in section 7.2 in this NASA Technical Standard and is not approved by the RFCB as appropriate for an optional approach as described in section 7.5 in this NASA Technical Standard shall comply with all of the following items:

a. Develop a DTA by complying with section 7.4.1 in this NASA Technical Standard.

b. Develop an IDMP by complying with section 7.4.2 in this NASA Technical Standard.

c. Develop an RTD by complying with section 7.4.3 in this NASA Technical Standard.

d. Develop loading spectra by complying with section 7.4.4 in this NASA Technical Standard.

e. Perform damage tolerance tests on coupons by complying with section 7.4.5 in this NASA Technical Standard.

f. Perform damage tolerance tests of hardware elements by complying with section 7.4.6 in this NASA Technical Standard.

g. Perform strength and life assessments by complying with section 7.4.7 in this NASA Technical Standard.

h. Perform damage tolerance tests of full-scale flight-like hardware by complying with section 7.4.8 in this NASA Technical Standard.

i.   Evaluate anomalies discovered during any portion of the BBA by complying with section 7.4.9 in this NASA Technical Standard.

[Rationale: Fracture critical parts need activities performed to understand the sensitivity of the part if a flaw or damage is present. These activities can range from a direct assessment of the part's capability with a flaw or damage to acceptance tests that establish the part has sufficient capability to a combination of activities that provides information deemed sufficient to mitigate the risk of failure caused by undiscovered flaws.]

Use of an alternative approach requires unique rationale and approval by the RFCB as described in section 10 [FCR 26] in this NASA Technical Standard. The approaches in this requirement are the preferred approaches if followed completely.

Damage tolerance assessment of composite or bonded hardware uses a BBA that includes testing, analysis, and certification. The testing includes material-allowable coupons, structural elements, subcomponents, components, and appropriate full-scale article testing. The tests are performed to evaluate relevant critical failure modes for loads that are representative of the hardware loading spectra and may include LEFs. The test elements develop assessment capability for credible damage levels as determined by the process steps resulting in the RTD. Such a BBA links multiple length scales and accounts for the effects of structural and material parameter variability.

Damage tolerance analysis of composite or bonded hardware is generally considered insufficiently developed to certify flight hardware without the support of a test program and the BBA. However, when a test-verified analysis approach exists and is applicable, an analysis approach that minimizes some of the testing detailed below may be submitted to the RFCB for consideration and approval. The assessment establishes that the spaceflight hardware meets all the criteria for life, strength, and damage tolerance detailed in these subsections. The details of the assessment are documented in the FCP.

The steps used in a damage tolerance assessment of composite or bonded hardware by incorporating the BBA and damage threat mitigation activities are detailed in the sections cited below:

a.   The initial three steps (sections 7.4.1 through 7.4.3 in this NASA Technical Standard) establish the critical damage states. There is likely an interaction between these three elements as flaw detection and impact damage protection/detection strategies are developed and implemented on the flight hardware. The final RTD is used in the certification of the flight hardware. Note that there may be credible damage conditions that occur at any point during service life, including during the mission.

b.   Concurrent with these first steps is development of the loading spectra determination (section 7.4.4 in this NASA Technical Standard) that affects the criticality of the remaining damage determined by the RTD.

c. The next four steps (sections 7.4.5 through 7.4.8 in this NASA Technical Standard) establish the structural response to the damage by both analysis and test at increasing levels of geometric complexity. There is also an interaction between these tests and the determination of critical damage states needed to develop the RTD.

d. Finally, discrepancies between the anticipated and observed test responses to damage initiation or growth are reconciled in accordance with section 7.4.9 in this NASA Technical Standard.

In practice, there will be iteration between and among these various steps.

BBA as described in this section is a comprehensive approach. Developers may have alternative approaches better suited to their hardware. These approaches and their rationale should be discussed with the RFCB.

7.4.1 Damage Threat Assessment

Develop a DTA according to the following to satisfy requirement [FCR 13] section 7.4.a in this NASA Technical Standard.

a. Provide information for residual strength sensitivity to impact damage and manufacturing flaws based on test data.

b. Define and quantify the flaws from any source that may occur to the hardware during its service life.

7.4.2 Impact Damage Mitigation Plan

Develop an IDMP according to the following to satisfy requirement [FCR 13] section 7.4.b in this NASA Technical Standard.

a. Define, document, and implement impact protection and/or detection strategies that are used for the flight hardware to diminish targeted damage threats identified by the DTA.

b. Prescribe when and how impact protection and/or detection strategies are to be used for flight hardware to mitigate credible damage or threats.

7.4.3 Residual Threat Determination

Develop an RTD according to the following to satisfy requirement [FCR 13] section 7.4.c in this NASA Technical Standard.

a. Define the worst-case credible flaw conditions that are shown to be tolerated by the hardware through analysis and test, considering all applicable flaw detection and mitigation strategies that are implemented for the flight hardware.

b. Encompass all possible worst-case credible damage conditions, except the threats that are mitigated by NDE evaluations, the IDMP, and the threats where risk is accepted by the program or project.

c. Document the damage states the program or project has chosen to exclude from the design.

The RTD helps identify flaws or damage conditions that are not screened by a combination of inspection, protection, and detection strategies.

Although inspection techniques meeting the 90 percent detectability level with 95 percent confidence called for in NASA-STD-5009 for metals are generally not available for composite or bonded materials, the RTD damage detection levels are to be set to produce a similar level of reliability as expected from metallic fracture critical parts.

For re-flight hardware, the inspections to be performed between flights are to be defined.

7.4.4 Loading Spectra

Establish that all the loads and the number of cycles or duration during the service life of the part at the appropriate environment are included to develop loading spectra to meet requirement [FCR 13] section 7.4.d in this NASA Technical Standard.

Development of the loading spectra includes all the applicable loads listed in section 7.3.1 in this NASA Technical Standard and all other applicable loads such as those related to environment effects on composite or bonded materials.

7.4.5 Damage Tolerance Tests of Coupons

Damage tolerance tests on coupons are performed with the applicable environments to generate a strength-based and a life-based database.

Perform damage tolerant coupon tests according to the following to satisfy requirement [FCR 13] section 7.4.e in this NASA Technical Standard:

a. Perform damage tolerance tests that represent flight hardware materials, manufacturing methods, and layups.

b. Perform damage tolerance tests that contain induced flaws and damage that encompass the worst-case credible-flaw conditions as determined by the RTD.

c. Perform damage tolerance tests that represent the modes of failure expected in the flight hardware.

d. Perform tests in a quantity sufficient to define design values for the relevant critical failure modes, e.g., residual strength, fatigue, using the B-basis statistical techniques as defined in CMH-17-1G or an equivalent approach approved by the RFCB.

e. Develop or use coupon data to establish the sensitivity of residual strength to impact and manufacturing damage as determined in the DTA in accordance with section 7.4.1 in this NASA Technical Standard.

Note that sufficient quantities of data are also necessary for use in computing the Weibull shape parameters used in determining the LEF, as described in CMH-17-1G.

7.4.6 Damage Tolerance Tests of Hardware Elements

Damage tolerance tests on hardware elements, subcomponents, and components are representative of the flight designs and have induced RTD determined flaws.

Perform damage tolerance tests of hardware elements according to the following to satisfy requirement [FCR 13] 7.4.f in this NASA Technical Standard:

a. Include both residual strength and life-based testing.

b. Perform tests sufficient in number to guide the design and provide confidence that the tests performed in accordance with section 7.4.8 in this NASA Technical Standard encompass the worst-case credible conditions, locations, and orientations.

Note that spectrum truncation is allowed for structural-level testing (components and full-scale hardware) with supporting coupon test data.

7.4.7 Strength and Life Assessments

Assessment of the flight article should be developed that is supported by analysis of the coupon and hardware element testing with RTD determined flaws present at any location and orientation.

Perform strength and life assessments according to the following to satisfy requirement [FCR 13] section 7.4.g in this NASA Technical Standard:

a. Perform analysis to establish that the B-basis residual strength after 1 service lifetime is sufficient to support DUL, after which the hardware will perform as intended.

b. Establish that the hardware performs as intended after experiencing a B-basis number of spectrum loading service lifetimes followed by one DLL cycle.

Note that the service life factor in analysis is the full B-basis number of lives, because the additional lives can be assessed without significant additional cost. One can therefore consider that no LEF is used or equivalently LEF=1.

7.4.8 Damage Tolerance Tests of Full-Scale Flight-Like Hardware

Perform damage tolerance tests of full-scale flight-like hardware according to the following to satisfy requirement [FCR 13] section 7.4.h in this NASA Technical Standard:

a. Induce flaws into test hardware as specified by the RTD in the worst credible location and orientation.

b. Perform NDE on test hardware before test to verify that the RTD flaws have been imposed and to record any flaws in addition to those imposed.

c. Account for the effects of environments and flight hardware structural conditions to simulate performances throughout the specified service lifetime. If tests are not performed in the operational environment, test levels are adjusted via an ECF.

d. Establish ultimate load capability in the test hardware after a minimum of 1 service lifetime loading.

e. Subject the test hardware to a minimum of 4 service lives of spectrum loading with appropriate LEF necessary to establish B-basis reliability followed by 1 DLL cycle.

More than 4 lifetimes of testing may be performed to reduce the LEF.

f. Establish that the test hardware does not experience structural failures and is capable of performing its design function after both spectrum service life testing and DUL testing (7.4.8.d and 7.4.8.e above).

(1) Determine primarily by assessment.

Functional or other tests may also be used. Note that items 7.4.8.a through f may be satisfied with one test article or may involve more than one test article as appropriate. The RFCB should be consulted for further understanding of what is expected to satisfy item 7.4.8.f, e.g., no structural failure or burst, no catastrophic leak caused by flaws, no catastrophic mechanical malfunctions.

(2) Perform NDE as part of this assessment.

7.4.9 Evaluate Flaws or Damage that Occur during BBA Testing

Evaluate flaws or damage occurring during BBA testing according to the following to satisfy requirement [FCR 13] section 7.4.i in this NASA Technical Standard:

a.   Evaluate unexpected flaws or damage, significant or unusual flaw growth, and any new failure modes observed.

b.   Address any concerns raised by the evaluation by assessment, test, retest, or redesign as appropriate.

c.   Include RFCB involvement with all assessments and evaluations.

7.5   Optional Approaches for Fracture Critical Parts

[FCR 14] Each fracture critical part that is not of a specific hardware type as described in section 7.2 in this NASA Technical Standard and is approved as appropriate for one of the following optional approaches by the RFCB shall comply with one of the following items:

a.   Single-event fracture critical components comply with section 7.5.1 in this NASA Technical Standard.

b.   HCF components comply with section 7.5.2 in this NASA Technical Standard.

c.   Proof test approach for composite or bonded hardware complies with section 7.5.3 in this NASA Technical Standard.

d.   Fleet leading testing approach complies with section 7.5.4 in this NASA Technical Standard.

e.   Hazardous fluid containers for payloads and experiments comply with section 7.5.5 in this NASA Technical Standard.

[Rationale: Parts that comply with this requirement have had sufficient activities performed to establish adequate risk mitigation of failure caused by the presence of a flaw or crack-like defect and are approved by the RFCB.]

Use of an alternative approach requires unique rationale and approval by the RFCB as described in section 10 [FCR 26] in this NASA Technical Standard.

7.5.1   Single-Event Fracture Critical Components

Fracture critical components with a single-event life loading history, such as pyrotechnic components, may be shown acceptable by demonstrating a factor of 1.4 on critical stress intensity factor instead of a factor of 4 on life, if all of the following conditions apply.

For single-event fracture critical components, satisfy the following items 7.5.1.a, 7.5.1.b, 7.5.1.c, and either 7.5.1.d or 7.5.1.e (as appropriate for the material and situation) to meet requirement [FCR 14] section 7.5.a in this NASA Technical Standard:

a. The single-event loading is a single cycle or a single cycle with rapidly decaying subsequent cycles.

b. The component is not subject to any other significant loads.

c. The evaluation, whether by analysis or testing, and any deviations from the prescribed approaches in this section are coordinated in advance with and approved by the RFCB.

d. Metallic components are shown by analysis to satisfy a minimum factor of 1.4 on critical stress intensity factor.

The margin is be computed as:

$$Margin\ on\ Critical\ Stress\ Intensity\ Factor = \frac{critical\ stress\ intensity\ factor}{(1.4 \times K_{applied})} - 1$$

where the:

> *critical stress intensity factor is usually represented as the plane strain fracture toughness, K_{Ic}, or a parameter such as K_{JIc} with approval of the RFCB.*

e. Both non-metallic components and metallic components satisfy the requirements of this section by using process controls that ensure the flight hardware will be represented by tests conducted on identical samples that establish the following:

Tests may be used in situations where the applied loads are difficult to determine, the material properties are uncharacterized, or other factors make the damage tolerance analyses difficult.

(1) Tests include a flaw in the worst location and orientation in the test articles.

(2) Apply either approach A or B below to establish the components are acceptable:

    A. Use this approach when loads are known and can be readily applied to test articles.

        i. The test load is at least 1.4 times the maximum expected flight load.

        ii. The flaw size is at least as large as the detectable sizes in NASA-STD-5009 (RTD for composite or bonded hardware, as described in section 7.4.3 in this NASA Technical Standard) for the inspection method applied to the flight hardware.

    B. Use this approach when loads are not well characterized or are difficult to apply.

i. The flaw size is at least twice as large in all dimensions as the detected sizes in NASA-STD-5009 (twice as large as the RTD for composite or bonded hardware as described in section 7.4.3 in this NASA Technical Standard) for the inspection method applied to the flight hardware.

ii. The load application is to simulate worst-case flight conditions.

iii. A sufficient number of articles are tested to ensure the test conditions approach the maximum flight conditions.

7.5.2   High-Cycle Fatigue Components

Fracture critical components operating in a potential HCF environment may be shown acceptable by establishing no HCF flaw growth. Examples of these are turbine blades, rotors, impellers, and other high-speed elements that are subject to local modes of high-frequency vibration and large numbers of loading cycles.

Satisfy the following for HCF Components to meet requirement [FCR 14] section 7.5.b in this NASA Technical Standard:

a. Use a value for fatigue crack growth threshold that has been approved by the RFCB.

b. Assume the initial NDE flaw size in the worst location and orientation.

c. Propagate the flaw (by analysis or test) for 4 times the required design life using the low-cycle loads.

d. Use the final flaw size from the calculations or test data in 7.5.2.c (above) as the initial flaw size in calculating the stress intensity factor (metallic components) or total strain (composite or bonded components) related to the HCF environment.

(1) The metallic component is acceptable if the calculated HCF stress intensity factor is below the stress intensity factor threshold for the metallic material.

(2) The composite or bonded component is acceptable if the calculated net section strain (or stress) is below the no-growth threshold strain (or stress) for the composite or bonded material with RTD determined flaws.

All items 7.5.2.a through 7.5.2.d are typically performed analytically. Items 7.5.2.b and 7.5.2.c may be performed by test.

7.5.3   Proof Test Approach for Composite or Bonded Hardware

Proof test, as an optional approach, is a category available on a limited-use basis. Use of this classification should include the RFCB early in the program. The proof test classification is usually limited to payload or secondary structures. These structures should have well defined load paths, loads, and boundary conditions. The proof test should adequately load all appropriate members and sections of the structure, where necessary both in tension and compression (load reversal). In cases where shear and/or compression dominate, the proof test approach may not be appropriate because of delamination growth under these load conditions. If proof test does not adequately replicate operational conditions, this may not be an applicable approach.

Satisfy the following for the proof test approach for composite or bonded hardware to meet requirement [FCR 14] section 7.5.c in this NASA Technical Standard:

a. Proof test the flight article to 1.2 times the limit load using one of the following:

(1) Conduct the proof test in the appropriate environment.

(2) Adjust the test loads using a coupon or hardware element test verified ECF.

b. Perform pre-proof and post-proof NDE, including special visual inspection if necessary, on the hardware.

c. Repair or replace hardware with indications of flaw growth or initiation that are discovered during proof test or with post-proof NDE.

(1) Repeat the proof test to 1.2 times the limit load for repaired hardware.

(2) Perform pre-proof and post-proof NDE, as well as special visual inspection if necessary, on the repaired regions.

d. Define the threats that may cause flaws from any source that may occur to the hardware during its service life, considering all applicable flaw detection and mitigation strategies that are implemented for the flight hardware.

e. Develop and implement an IDMP for the hardware that assures a complete record of hardware impact or damage status and mitigates the risk of undetected damage from the threats identified in 7.5.3.d (above) for the period between post-proof NDE and launch.

f. Establish that the largest remaining residual threat after post-proof NDE through the remainder of the service life can create damage no larger than the flaw size screened by NDE.

g. Repeat the proof test, repair, or replace the hardware as described in 7.5.3.a through 7.5.3.c (above) if any incidents of impact or other damage occur after post-proof NDE and before launch.

h. For re-flight hardware, repeat the proof test approach activities in items 7.5.3.a through 7.5.3.g in this NASA Technical Standard before the hardware is re-flown.

Proof test loads should be limited to less than 80 percent of ultimate strength of the structure for the appropriate mode of failure, e.g., tension, compression, and shear. Structures with an ultimate safety factor of 1.5 or greater will preclude exceeding 80 percent of ultimate strength when using a test factor of 1.2. Note that the full DTA activities of section 7.4.1 in this NASA Technical Standard are not required. However, test data describing capability relative to damage or flaws will likely be necessary to assist with disposition of any flaws discovered during pre-proof NDE. Test data for capability relative to damage or flaws may also be necessary to develop NDE criteria for reportable flaws. The relevant capability is dependent on the failure mode of concern, e.g., compression-after-impact strength, delamination growth, or other. Proof tests are usually performed in the operational environment, or the test levels are adjusted via an ECF.

The composite or bonded structure should be designed so that accompanying metallic parts do not experience detrimental yielding during the proof test.

7.5.4   Fleet Leader Testing

Satisfy the following for fleet leader testing to meet requirement [FCR 14] section 7.5.d in this NASA Technical Standard:

    a. Provide the approach and rationale to the RFCB for approval before implementation.

    b. Document the approved approach in the FCP.

In cases where loading conditions are poorly defined, a ground test fleet leader program that allows use of the hardware may be feasible.

7.5.5   Hazardous Fluid Containers for Payloads and Experiments

The hazardous fluid containers category is limited to payload and experiment applications at conditions defined in requirements below. This hardware type is not part of a pressurized system nor is it intended to transfer stored fluid as part of a pressurized system.

Satisfy the following for hazardous fluid containers for payloads and experiments to meet requirement [FCR 14] section 7.5.e in this NASA Technical Standard:

    a. The container is limited to an MDP of 152 kPa (22 psi, 1.5 atm) and a maximum volume of 0.05 m^3 (1.76 ft^3).

    b. An analysis is to show a positive margin against burst when a factor of 2.5 on MDP is used.

    c. Perform proof test to 1.5 MDP.

    d. Establish that no damage or detrimental deformation exists after the proof test.

e. Establish damage tolerance against rupture and leak by satisfying sections 8 and 9 in this NASA Technical Standard for all materials, section 7.3 in this NASA Technical Standard for metallic parts, section 7.4 in this NASA Technical Standard for composite or bonded parts, and by test or analysis as approved by the RFCB for other materials.

f. In addition to section 8 requirements in this NASA Technical Standard, perform an NDE inspection of all fusion joints in the container after proof test to determine acceptable conditions both on the surface and within the fusion joint.

g. Perform a leak test to 1.0 times the MDP.

In instances where NDE is not feasible, the manufacturer may employ a process-control program that assures the quality of the uninspectable welds and obtain approval of the RFCB.

Proof tests are usually performed in the operational environment, or the test levels are adjusted via an ECF.

Inertial load effects (including attach points) may necessitate additional assessments beyond the items in this category.

8. FLAW SCREENING, TRACEABILITY, AND MATERIAL SELECTION

[FCR 15] All fracture critical parts shall be screened for flaws with methods and techniques identified in the FCP.

[Rationale: An understanding of the flaws or damage types to be screened and the methods to be used is necessary to assure adequate fracture control implementation.]

NDE is the primary method used for screening flaws for fracture critical parts. Proof test of the flight article may be used to screen for flaws in special cases, especially for glass elements. Visual inspection is an NDE method that is frequently used for inspecting composite or bonded parts for damage, in addition to other NDE methods. Visual inspection is also used for inspecting optical elements for flaws, often in addition to proof testing. In some cases, process control may be allowed as a method for establishing an upper bound on flaw sizes that may be present in the part.

8.1 Flaw Screening

8.1.1 NDE for Metallic Parts

[FCR 16] Metallic fracture critical parts screened with NDE shall have inspections performed in accordance with NASA-STD-5009 and include the following for flaw screening by NDE:

a. Apply sufficient flaw inspection methods to the flight hardware to screen flaws larger than or equal to the size and shape that are evaluated in the hardware damage tolerance assessment.

b. In addition to NDE for flaw screening of other regions of fracture critical parts, perform post-proof test NDE at critical welds and other critical locations identified in the FCP for all parts that are proof tested as a part of acceptance, i.e., critical hardware locations not screened for specific flaws with the proof test.

[Rationale: This cites NASA-STD-5009 and reduces the potential for redundant or conflicting requirements.]

It is expected that fracture critical parts have surface and volumetric inspections unless there is rationale that it is not necessary. The need for internal (volumetric) inspection depends on application and materials characteristics such as thickness, product form, and other factors. Internal inspection requirements and methods should be determined early in the design process so that proper flaw screening is accomplished.

According to NASA-STD-5009, the flaw sizes and shapes that are evaluated in the hardware damage tolerance assessment are based on 90-percent probability of detection with 95-percent confidence (90/95 or better) flaw detection capability.

If one NDE method cannot adequately examine a part, additional NDE methods may be needed. If there are multiple types of flaws or complex geometry to assess, additional NDE may be needed. If there is uncertainty about which NDE methods or results for a particular part are to be used to define flaws for the damage tolerance assessment, conservative choices are to be made.

NDE activities and damage tolerance assessment activities should be coordinated to assure flaw screening occurs in the way intended.

8.1.2 NDE for Composite or Bonded Parts

NDE activities for composite or bonded materials requirements apply to fracture critical and NFC parts. Because of the potential sensitivity to impact damage and flaws for these types of materials, additional activities are necessary for NFC parts in accordance with [FCR 9] 6.3.c in this NASA Technical Standard.

[FCR 17] For composite or bonded materials, the hardware developer shall:

a. Provide the NDE methodology and rationale in the FCP.

b. Perform flaw screening by NDE on all composite or bonded part regions, except for the following:

(1) No NDE is required for NFC low-released mass parts.

(2) No NDE is required for NFC contained parts.

    c. For hardware that is proof tested as part of acceptance, perform pre-proof and post-proof test NDE at critical joints, discontinuities, and other critical locations identified in the FCP for all hardware, i.e., critical hardware locations not screened for specific flaws with the proof test.

[Rationale: There are no NDE standards available that are applicable to the wide variety of non-metallic materials and forms in use and the different NDE methods required for their inspection. The approach for NDE of other materials needs to be documented and fully explained within the FCP.]

Inspection of composite or bonded parts is to meet the intent of MIL-HDBK-6870, Nondestructive Inspection Program Requirements for Aircraft and Missile Materials and Parts, as required in NASA-STD-6016.

Hardware should receive post-proof NDE unless a special RFCB approval has been granted.

Generally, the NDE approach and rationale for all materials should address which indications rise to the level of a reportable flaw. For signal-based methods, such as ultrasonic inspections, NDE acceptance criteria are usually necessary to discern whether the signal responses warrant nonconformance reporting. All damage indications from visual inspection are reportable. Workmanship standards for visual inspection should define acceptance criteria, e.g., porosity, surface texture, geometric contours. NDE acceptance criteria may be developed by analysis with supporting coupon test data for the appropriate material type. Prior approval should be obtained from the RFCB when visual inspection is used as a flaw-screening technique for fracture control. Screening of a low-risk part with NDE should be considered when it is plausible for that part to be reclassified as a fracture critical part. A part may need to be reclassified when it is plausible for that part to be accepted for flight with out-of-tolerance dimensions or nonstandard material properties.

8.1.3 Proof Test

[FCR 18] If proof testing is used as the flaw screening technique for fracture critical parts, the approach shall be documented in the FCP with rationale establishing that it is an applicable approach that has been approved by the RFCB.

[Rationale: Proof test may be used for flaw screening. However, few parts, materials, and applications lend themselves to a simple proof test strategy. Environmental effects, temperature, test fixture, inertial loads, and other complexities require careful consideration before accepting proof as the sole method for flaw screening. If proof test is used for flaw screening, an understanding of the planned approach and anticipated effectiveness needs to be approved by the RFCB and documented in the FCP.]

Proof test should not be used as the only flaw screening method for composite or bonded hardware.

The flaw size used in the life assessment should adequately account for flaw growth during the proof test. To establish that the assessment is valid, sufficient test data should be obtained using pre-flawed specimens that are representative of the part configuration, material conditions, and screened flaw and show the amount of growth of all crack fronts during the proof test from all sources, including stable tearing, and both EAC and SLC if applicable, have been conservatively bounded.

When it is judged that a proof test is appropriate to screen hardware for flaws, the proof test should occur at the in-service temperature and environment. If this is not feasible, an ECF can be used as approved by the RFCB. Upper bound critical stress intensity or residual strength should be used when establishing an analytically predicted flaw size screened by proof test.

Note that a proof test is required for acceptance in accordance with NASA-STD-5001 (or program-specific requirements), with a minimum proof test factor, depending upon whether a prototype or proto-flight verification approach is followed and the type of material used.

8.1.4   Process Control

[FCR 19] If process controls are used to establish bounds on flaw sizes in fracture critical parts, the approach shall be subject to the following:

    a. The approach is documented in the FCP.
    b. The rationale establishing that the approach is applicable is documented in the FCP.
    c. The FCP is approved by the RFCB.

[Rationale: Use of process control information to define flaws or damage that could be in the part is an unusual approach. An understanding of the approach and supporting information need to be approved by the RFCB and documented in the FCP.]

Process control rationale to bound flaw sizes submitted for RFCB approval should include documentation on why this approach is being applied, an overview of the hardware, and evaluation that the approach is adequate for fracture control. Descriptions of the relevant manufacturer's experience base, process control during manufacture, inspection results, and subsequent life of the component, all component testing, and summary arguments should be included.

8.1.5   Detected Flaws

[FCR 20] Spaceflight hardware with detected flaws that is used for flight without being repaired or replaced shall have a specific detailed assessment approach documented with rationale in the FCP that contains the following:

a. An assessment approach of metallic parts by analysis is to include the following items in addition to the items in section 7.3.2 in this NASA Technical Standard:

(1) Upper bound flaw size.
(2) Upper bound crack growth rate.
(3) Lower bound critical stress intensity factor or residual strength.
(4) Lower bound cyclic fatigue crack growth threshold stress intensity range (ΔK_{th}).

b. An assessment approach for composite or bonded parts with detected flaws is to include the following items:

(1) The approach and rationale provided to the RFCB for approval before implementation.

(2) Documentation of the approved approach in the FCP.

[Rationale: An understanding of the approach and methodology to accept detected flaws, which accounts for variability in the assessment, is necessary to assure adequate fracture control implementation.]

For reportable detected flaws in composite or bonded parts, a similar worst-case analysis approach to that used for metal parts may not be available. Any proposed analysis approach is to be test verified with a similar damage configuration and approved by the RFCB.

Note that the detailed assessment approach may be by damage tolerance test if approved by the RFCB.

The normal fracture control process is carried out with the assumption that the part contains a flaw in the worst-case location and orientation. The assessment of the assumed flaw includes typical fracture properties and an assumed flaw size. However, when flaws are detected in a part that is planned for use in flight, an assessment is performed using bounding flaw sizes, material properties, loads, and boundary conditions.

Fracture critical parts with reportable NDE indications are to be assessed by a process approved by the Technical Authority to determine whether the indication is a flaw.

Fracture critical parts with detected flaws are to be assessed with an approach that satisfies [FCR 20] and then evaluated by the Technical Authority to determine whether the part is acceptable to use as is or if the part is to be repaired or replaced. If the part is to be repaired, the repair process is to be an established, proven process that has been approved for this purpose by the Technical Authority.

Pressure vessels and COPVs should not be flown with detected flaws, e.g., see sections 7.2.1 and 7.2.2 in this NASA Technical Standard. If an exception is sought, it is to satisfy section 10 in this

NASA Technical Standard as an alternative approach that is deviating from these established procedures and needs approval by the Technical Authority and the RFCB.

The RFCB should be notified of the intent to fly the flawed part when it is not feasible to repair or replace the part.

8.2   Traceability for Fracture Control

Traceability requirements are typically associated only with fracture critical parts, except in the case of NFC composite or bonded parts. Because of the nature of these types of materials, additional activities are necessary for NFC parts in accordance with [FCR 9] section 6.3.d in this NASA Technical Standard.

[FCR 21] Traceability for each fracture critical and NFC composite or bonded part shall be established and maintained by providing a unique serial number (or other method when serialization is not practical) and a complete life history, including load history, impact damage, repair, materials, manufacturing, processing, and environmental exposure.

[Rationale: Traceability is necessary to assure the information used to assess flaw or damage sensitivity, screening, and protection is understood and accurate throughout the service life of the hardware.]

Traceability for NFC composite or bonded parts is somewhat unique relative to metallic parts. While metallic parts usually have a specification for providing minimum properties throughout the part, composite and bonded parts are composed of elements that may have specifications, but the properties after combination of these elements are often unique to the part being produced. These considerations lead to the need for traceability of fracture critical parts and NFC composite or bonded parts as also required in section 6.3 in this NASA Technical Standard.

8.3   Material Selection and Usage for Fracture Critical Parts

Material selection and usage requirements are typically associated with fracture critical parts, except in the case of NFC composite or bonded parts. Because of the nature of these types of materials, additional activities are necessary for NFC in accordance with [FCR 9] 6.3.f in this NASA Technical Standard.

[FCR 22] The selection, processing, and use of materials for all fracture critical and NFC composite or bonded parts shall include the following items, which are documented directly in the FCSR or the items have pertinent documents referenced in the FCSR:

   a.   Fabricate parts from materials with supplier data certifications.

   b.   Select materials compatible with NASA-approved Standards and Specifications.

   c.   Account for the effect of operating conditions on damage tolerance properties.

Examples of conditions that may affect damage tolerance properties are temperature, operating environment (atmosphere, corrosive media), cleaning and/or inspection agents, coatings, proof test fluids, loading spectra, time, temperature, and other environmental exposures and conditions.

d. Design and assess with strength and damage tolerance properties that are generated by tests on samples representative of the flight hardware material, subject to either item (1) or (2) below:

(1) Material is processed to the same thickness, material process condition, and material orientation in the part that result in the worst combination for damage tolerant assessment.

(2) The material process condition and the material orientation are fully traceable throughout fabrication and service life.

Examples of activities that may affect a metallic material process condition include: mill billet hot processes, such as forging, rolling, or other high-deformation processes; metallurgical product operations, including heat treatments; shaping operations, such as rolling, spinning, or drawing; fabrication joining processes, such as welding; and any other operations known to affect the material microstructure, strength, fracture, crack growth, or environment sensitivity properties.

e. Derived strength and damage tolerance data obtained from NASA-approved sources. If data are lacking, data are conservatively bounded or determined by sufficient testing to assess scatter to provide averages with testing approved by the RFCB.

f. Obtain an approved MUA for any materials not developed and qualified in accordance with the requirements of NASA-STD-6016.

g. Include all MUAs in the FCSR.

[Rationale: The specific items related to materials selection and usage are necessary to assure the information used to assess flaw or damage sensitivity is understood and accurate throughout the service life of the hardware.]

9. FRACTURE CONTROL DOCUMENTATION AND VERIFICATION

9.1 Fracture Control Documentation

9.1.1 Fracture Control Plan

The FCP developed in compliance of [FCR 1] in section 4.1 in this NASA Technical Standard is part of the documentation.

9.1.2 Engineering Drawings

[FCR 23] Fracture critical parts shall be identified on engineering drawings in the notes of the individual part drawing, along with the inspection, serialization, and other pertinent information necessary to maintain traceability of the part and its history of manufacturing and use.

[Rationale: Identification of fracture critical parts on engineering drawings is necessary to assure that the appropriate NDE, serialization, and traceability needs are recognized and implemented.]

The type of NDE and NDE acceptance criteria should be specified.

Detected flaws are assessed in accordance with section 8.1.5 in this NASA Technical Standard.

Processing or fabrication requirements that would affect fracture properties of a fracture critical part in a given application, such as heat treatments, welding requirements, and peaking/mismatch allowables, grain or fiber direction, and other critical parameters, should be specifically called out on the part drawing.

Composite or bonded material epoxies and adhesives should have their shelf life requirements included as part of the engineering drawing notes.

9.1.3 Fracture Control Summary Report

[FCR 24] An FCSR shall be developed by the spaceflight hardware program or project that:

    a.   Documents the basis for acceptance that all the flight hardware parts have met the fracture control requirements in the approved FCP.

    b.   Contains detailed information or reference to detailed information for all parts, including results for evaluations, classification, assessments, inspections and other pertinent records, and their disposition for fracture.

    c.   Documents all assessments, such as analyses and tests, conducted on representative flight hardware used for flight certification.

    d. Identifies the flaws and impact damage threats that are accepted on risk by the program authority, i.e., the flaws and impact damage threats for which there is no damage tolerance evaluation.

    e. Is approved by the RFCB.

[Rationale: The FCSR contains the information or summarizes and points to the detailed reports necessary to show fracture control compliance of all parts to the requirements in the approved FCP.]

The FCSR may point to other project documentation that is available for review by the RFCB that contains fracture control data relevant to the completion of the FCSR as necessary to avoid duplication of efforts.

The flaws identified in 9.1.3.d (above) may vary from program to program. Examples may include flaws, such as those caused by lightning strikes, system failures, handling mishaps, MMOD impacts, bird impacts, etc.

9.1.3.1   Detailed Information for the FCSR

a.   *The FCSR provides sufficient information to certify that fracture control requirements have been met by assessment results available in detailed damage tolerance assessment reports of analyses and testing. The FCSR summarizes the results and damage tolerance service life, loadings, flaw screening methods, initial flaw sizes used in the assessment, material characteristics, flaw sizes at the end of lifetime, predicted lifetime, and analysis methods used in the assessment.*

b.   *The FCSR provides sufficient hardware descriptions, including sketches and figures, to convey a clear understanding of the hardware elements and their functions.*

c.   *Supporting detailed documentation, such as drawings, calculations, analyses, testing details, test results, data printouts, inspection plans, records, DTA, IDMP, RTD, specifications, certifications, MUAs, reports, procedures, and all other items that establish the fracture control suitability of the flight hardware, is to be provided to the RFCB under separate cover, upon request.*

d.   *The FCSR gives an accounting of all parts and their disposition for fracture control as follows:*

(1) *Identifies exempt parts, groups of exempt parts, and types of exempt parts.*

(2) *Lists NFC parts, along with their classification and supporting rationale.*

(3) *Lists fracture critical parts with a summary of the basis for their damage tolerance.*

e.   *The FCSR identifies the following for all NFC parts requiring NDE, including fail-safe parts, containment enclosures, NHLBB items, low-risk parts, NFC composite or bonded hardware assessed in accordance with section 6.3.1 in this NASA Technical Standard, and fracture critical parts:*

(1) NDE and other inspections carried out on the parts.
(2) MUAs.

f. The FCSR identifies inspections and other requirements imposed on re-flight hardware before re-flight.

g. The FCSR includes results from implementation of approved alternative approaches used in accordance with section 10 in this NASA Technical Standard.

9.1.3.2 Other Documentation

Other documentation supporting fracture control may be called for in the program data requirements. A summary of any parts with known flaws that were accepted for flight by the Technical Authority and any accompanying RFCB review documentation should be maintained by the program/project configuration data management organization. This includes any discrepancies or deviations from design that affect fracture control, e.g., any flaw detection information with resolution data.

9.2 Verification

[FCR 25] Verification of adherence of the flight hardware to the fracture control requirements in this NASA Technical Standard shall include all of the following:

a. Written documentation that establishes that each requirement has been met. This documentation describes how the requirement was verified, e.g., test, analysis, inspection. The project is responsible for providing this verification, including assurance that fracture control activities were implemented on the flight hardware before flight and reflight, to the appropriate program management.

b. Approval of the FCP and FCSR by the RFCB, documented by a concurrence memorandum from the RFCB to the applicable project/program office.

c. In the event of conflict between the RFCB and the applicable project office concerning verification of compliance with fracture control requirements, follow the procedures in place at each NASA Center to resolve technical conflict, with the option to appeal to the NASA Chief Engineer for final resolution.

[Rationale: All requirements need to be verified. The verification is the evaluation and documentation that all requirements have been met. There are many methods of verification, e.g., analysis, test, inspection, each of which should be documented.]

The project is responsible to the appropriate program for the line-by-line review of the verification requirements. The RFCB is responsible for a review of the methodology of the compliance to and verification of the requirements. These are documented in the FCP and FCSR, respectively.

Commonly, the project writes project-specific requirements that are traceable to a higher level standard. This requirements set includes a verification section. In this section, there is a verification requirement for every requirement in the project-specific requirements. This requirement documents how the verification will be done.

The RFCB is to receive the FCP and FCSR in accordance with requirements in this NASA Technical Standard, but, as described above, the RFCB is not the entity that performs the requirement-by-requirement review.

10.   ALTERNATIVES

[FCR 26] If alternative approaches are proposed (rather than meeting any part of the accepted approaches that are prescribed in sections 5, 6, 7, or 8 in this NASA Technical Standard, with the exclusions shown below), the alternative approach shall include all of the following items:

a.   Provide an equivalent assurance of mitigating the risk of catastrophic failure from flaws during the service life of the hardware.

b.   Have the approval of the RFCB.

c.   Meet all the other applicable requirements in this NASA Technical Standard.

d.   FCRs 10, 15, 20, 21, and 22 (sections 7.1, 8, 8.1.5, 8.2, and 8.3, respectively, in this NASA Technical Standard) are excluded from alternative approach consideration.

Note that FCR 26 pertains to sections 5, 6, 7, and 8 only; therefore, FCRs 1, 2, 3, 4, 5, 23, 24, and 25 are also excluded from alternative approach consideration.

[Rationale: Regardless of the detailed acceptance approach used for each part, the method is to be responsive to NASA NPR directives to mitigate the risk of catastrophic failure related to flaws during the service life of the hardware. If an alternative approach is proposed, its effectiveness is to be established and the approach documented in the approved hardware-specific FCP in accordance with [FCR 1] in section 4.1 in this NASA Technical Standard.

This document contains acceptable methods for fracture control based on NASA's experience base, established approaches, industry standards, or aerospace standards that satisfy the fracture control requirements; therefore, it is advisable to use the methods prescribed in this NASA Technical Standard.

Alternatives to the approaches prescribed in this NASA Technical Standard in sections 5, 6, 7, or 8 may be proposed for a specialized part or application for which the approaches in this NASA Technical Standard are not feasible or effective or for which other viable methods are advantageous. Alternative approaches, accompanied by supporting rationale that establishes that the alternative has comparable rigor to the approaches in this NASA Technical Standard, are presented to the RFCB for review and approval. Approved alternative approaches are to be documented in the specialized FCP for the parts in accordance with [FCR 1] in section 4.1 in this NASA Technical Standard.

APPENDIX A

REQUIREMENTS COMPLIANCE MATRIX

A.1   Purpose and/or Scope

The purpose of this appendix is to provide guidance in the form of an easy look-up list of the fracture control requirements in this NASA Technical Standard. A listing of requirements is provided for selection and verification of requirements by programs and projects. *(Note: Enter "Yes" to describe the requirement's applicability to the program or project; or enter "No" if the intent is to tailor, and enter how tailoring is to be applied in the "Rationale" column.)*

A.2   Requirements Compliance Matrix Tables

Table 2, General Requirements, contains the general requirements for an FCP, the responsible parties for fracture control, and the fracture control classification of all parts. Table 3, Exempt, contains the requirements for the exempt classification. Table 4, NFC Requirements, contains the requirements for the non-fracture critical classification. Table 5, Fracture Critical Requirements, contains the requirements for the fracture critical classification. Table 6, Flaw Screening, Evaluation, and Materials Requirements, contains the flaw screening, evaluation, traceability, and material selection requirements. Table 7, Documentation and Verification, contains the requirements for documentation and verification. Table 8, Alternate Approach Requirement, presents the requirement for alternate approaches.

Table 2—General Requirements

		NASA-STD-5019A		
Description	**Section**	**Requirement in this Standard**	**Applicable (Yes or No)**	**If No, Enter Rationale**

Overarching fracture control requirement	4.1	[FCR 1] A Fracture Control Plan shall be developed and maintained by the program for human-rated spaceflight hardware that satisfies all of the following:
		    a. Addresses all of the parts in the program-specific flight hardware.
		    b. Meets the requirements of this NASA Technical Standard.
		    c. Specifies fracture controls that are established to mitigate the risk of catastrophic failure caused by flaws throughout the service life of the hardware
		    d. Has approval by the RFCB.		

NASA's implementation of fracture control on human-rated spaceflight hardware	4.2.1	[FCR 2] The NASA Center responsible for the human-rated spaceflight hardware shall establish and designate a NASA RFCB to ensure compliance with the technical requirements of this document.		
	4.2.2	[FCR 3] Human-rated spaceflight programs shall impose fracture control on their projects to meet the requirements of this NASA Technical Standard.		
	4.2.3	[FCR 4] Fracture control implementation shall be performed with the oversight, advice, and approval of the RFCB.		

Evaluation of all parts	4.3	[FCR 5] All parts used in human-rated spaceflight hardware shall be evaluated to identify the following:
		    a. The fracture control classification of each part as either exempt, NFC, or fracture critical.
		    b. The corresponding approaches that follow the requirements of this NASA Technical Standard to be documented in the FCP.		

Table 3—Exempt

NASA-STD-5019A				
Description	Section	Requirement in this Standard	Applicable (Yes or No)	If No, Enter Rationale
Exempt Classification	5	[FCR 6] Each part classified as exempt shall fit into one of the following categories: a. Non-structural parts with no credible failure mode caused by a flaw. b. Non-structural parts with no credible potential for causing a catastrophic hazard. c. Other non-structural parts approved by the RFCB for exempt status		

Table 4—NFC Requirements

Description	Section	Requirement in this Standard	Applicable (Yes or No)	If No, Enter Rationale
		NASA-STD-5019A		
Non-Fracture Critical Classification	6.1	[FCR 7] To be classified as NFC, each part that is described by a specific hardware type in the following list shall comply with the established approach given in the referenced subsection: a. NFC metallic fasteners, rivets, shear pins, and locking devices comply with section 6.1.1 in this NASA Technical Standard. b. NFC shatterable components and structures comply with section 6.1.2 in this NASA Technical Standard. c. NFC rotating hardware complies with section 6.1.3 in this NASA Technical Standard. d. NFC sealed containers comply with section 6.1.4 in this NASA Technical Standard. e. NFC tools, mechanisms, and tethers comply with section 6.1.5 in this NASA Technical Standard. f. NFC batteries comply with section 6.1.6 in this NASA Technical Standard.		
	6.2	[FCR 8] Each part classified as NFC that is not of a specific hardware type as described in section 6.1 in this NASA Technical Standard shall comply with one of the following items: a. NFC low-released mass complies with section 6.2.1 in this NASA Technical Standard. b. NFC contained complies with section 6.2.2 in this NASA Technical Standard. c. NFC fail-safe complies with section 6.2.3 in this NASA Technical Standard.		

		NASA-STD-5019A		
Description	**Section**	**Requirement in this Standard**	**Applicable (Yes or No)**	**If No, Enter Rationale**

		d. NFC NHLBB pressurized components comply with section 6.2.4 in this NASA Technical Standard.
		    e. NFC low-risk part complies with section 6.2.5 in this NASA Technical Standard.
		    f. NFC documented non-hazardous failure mode complies with section 6.2.6 in this NASA Technical Standard.		

	6.3	[FCR 9] NFC composite or bonded parts that satisfy requirements for classification in a specific category in sections 6.1 and 6.2 in this NASA Technical Standard shall also comply with all of the following items:
		    a. For parts classified as NFC low risk, develop the following:
		        (1) A DTA in accordance with section 7.4.1 in this NASA Technical Standard.
		        (2) An IDMP in accordance with section 7.4.2 in this NASA Technical Standard.
		        (3) An RTD in accordance with section 7.4.3 in this NASA Technical Standard.
		    b. For NFC parts not classified as low risk, perform the following:
		        (1) Define and quantify the flaws from any source that may occur to the hardware during its service life, considering all applicable flaw detection and mitigation strategies that are implemented for the flight hardware.
		        (2) Develop an IDMP in accordance with section		

		NASA-STD-5019A		
Description	**Section**	**Requirement in this Standard**	**Applicable (Yes or No)**	**If No, Enter Rationale**
		7.4.2 of this NASA Technical Standard. c. Perform NDE after completion of all manufacturing processes (or (or after proof test, if a proof test is performed) in accordance with section 8.1.2 in this NASA Technical Standard, with the following clarifications: (1) No NDE is required for NFC low-released mass parts. (2) No NDE is required for NFC contained parts. d. Meet the traceability requirement of section 8.2 in this NASA Technical Standard [FCR 21]. e. Meet the material selection and usage requirement of section 8.3 in this NASA Technical Standard [FCR 22].		

Table 5—Fracture Critical Requirements

	NASA-STD-5019A			
Description	**Section**	**Requirement in this Standard**	**Applicable (Yes or No)**	**If No, Enter Rationale**
Fracture Critical Classification	7.1	[FCR 10] Parts shall be classified as fracture critical unless one of the following is met: a. There is no credible possibility for a flaw in the part to cause failure during the lifetime of the part. b. Part failure does not result in a credible catastrophic hazard.		
	7.2	[FCR 11] Each fracture critical part that is described by a specific hardware type in the following list shall comply with the established approach given in one of the following items: a. Fracture critical metallic pressure vessels comply with section 7.2.1 of this NASA Technical Standard. b. Fracture critical composite overwrapped pressure vessels (COPVs) and composite overwrapped pressurized fluid containers comply with section 7.2.2 in this NASA Technical Standard. c. Other fracture critical pressure vessels and pressurized fluid containers comply with section 7.2.3 in this NASA Technical Standard. d. Fracture critical lines, fittings, and other pressurized components comply with section 7.2.4 in this NASA Technical Standard. e. Fracture critical habitable structures and volumes comply with section 7.2.5 in this NASA Technical Standard. f. Fracture critical pressurized structures comply with section 7.2.6 in this NASA Technical Standard.		

NASA-STD-5019A				
Description	Section	Requirement in this Standard	Applicable (Yes or No)	If No, Enter Rationale
		g. Fracture critical rotating hardware complies with section 7.2.7 in this NASA Technical Standard. h. Fracture critical fasteners comply with section 7.2.8 in this NASA Technical Standard. i. Fracture critical shatterable components and structures comply with section 7.2.9 in this NASA Technical Standard. j. Fracture critical tools, mechanisms, and tethers comply with section 7.2.10 in this NASA Technical Standard. k. Fracture critical batteries comply with section 7.2.11 in this NASA Technical Standard.		

	7.3	[FCR 12] Each fracture critical metallic part that is not of a specific hardware type as described in section 7.2 in this NASA Technical Standard and is not approved by the RFCB as appropriate for an optional approach as described in section 7.5 in this NASA Technical Standard shall comply with one of the following item combinations: a and b; a and c; or a, b, and c.
		  a. Develop loading spectra by complying with section 7.3.1 in this NASA Technical Standard.
		  b. Perform assessment by analysis to comply with section 7.3.2 in this NASA Technical Standard.
		  c. Perform assessment by test to comply with section 7.3.3 in this NASA Technical Standard.		

	7.4	[FCR 13] Each fracture critical composite or bonded part that is not of		

		NASA-STD-5019A		
Description	**Section**	**Requirement in this Standard**	**Applicable (Yes or No)**	**If No, Enter Rationale**
		a specific hardware type as described in section 7.2 in this NASA Technical Standard and is not approved by the RFCB as appropriate for an optional approach as described in section 7.5 in this NASA Technical Standard shall comply with all of the following items: a. Develop a DTA by complying with section 7.4.1 in this NASA Technical Standard. b. Develop an IDMP by complying with section 7.4.2 in this NASA Technical Standard. c. Develop an RTD by complying with section 7.4.3 in this NASA Technical Standard. d. Develop loading spectra by complying with section 7.4.4 in this NASA Technical Standard. e. Perform damage tolerance tests on coupons by complying with section 7.4.5 in this NASA Technical Standard. f. Perform damage tolerance tests of hardware elements by complying with section 7.4.6 in this NASA Technical Standard. g. Perform strength and life assessments by complying with section 7.4.7 in this NASA Technical Standard. h. Perform damage tolerance tests of full-scale flight-like hardware by complying with section 7.4.8 in this NASA Technical Standard. i. Evaluate anomalies discovered during any portion of the BBA by complying with section 7.4.9 in this NASA Technical Standard.		
	7.5	[FCR 14] Each fracture critical part that is not of a specific hardware type		

		NASA-STD-5019A		
Description	**Section**	**Requirement in this Standard**	**Applicable (Yes or No)**	**If No, Enter Rationale**
		as described in section 7.2 in this NASA Technical Standard and is approved as appropriate for one of the following optional approaches by the RFCB shall comply with one of the following items: a. Single-event fracture critical components comply with section 7.5.1 in this NASA Technical Standard. b. HCF components comply with section 7.5.2 in this NASA Technical Standard. c. Proof test approach for composite or bonded hardware complies with section 7.5.3 in this NASA Technical Standard. d. Fleet leading testing approach complies with section 7.5.4 in this NASA Technical Standard. e. Hazardous fluid containers for payloads and experiments comply with section 7.5.5 in this NASA Technical Standard.		

Table 6—Flaw Screening, Evaluation, and Materials Requirements

NASA-STD-5019A				
Description	**Section**	**Requirement in this Standard**	**Applicable (Yes or No)**	**If No, Enter Rationale**
Flaw screening and evaluation, traceability, and material requirements for fracture critical parts and other applicable components	8	[FCR 15] All fracture critical parts shall be screened for flaws with methods and techniques identified in the FCP.		
	8.1.1	[FCR 16] Metallic fracture critical parts screened with NDE shall have inspections performed in accordance with NASA-STD-5009 and include the following for flaw screening by NDE: a. Apply sufficient flaw inspection methods to the flight hardware to screen flaws larger than or equal to the size and shape that are evaluated in the hardware damage tolerance assessment. b. In addition to NDE for flaw screening of other regions of fracture critical parts, perform post-proof test NDE at critical welds and other critical locations identified in the FCP for all parts that are proof tested as a part of acceptance, i.e., critical hardware locations not screened for specific flaws with the proof test.		
	8.1.2	[FCR 17] For composite or bonded materials, the hardware developer shall: a. Provide the NDE methodology and rationale in the FCP. b. Perform flaw screening by NDE on all composite or bonded part regions, except for the following: (1) No NDE is required for NFC low-released mass parts. (2) No NDE is required for NFC contained parts. c. For hardware that is proof tested as part of acceptance, perform pre-proof and post-proof test NDE at critical joints, discontinuities, and		

		NASA-STD-5019A		
Description	**Section**	**Requirement in this Standard**	**Applicable (Yes or No)**	**If No, Enter Rationale**
		other critical locations identified in the FCP for all hardware, i.e., critical hardware locations not screened for specific flaws with the proof test.		
	8.1.3	[FCR 18] If proof testing is used as the flaw screening technique for fracture critical parts, the approach shall be documented in the FCP with rationale establishing that it is an applicable approach that has been approved by the RFCB.		
	8.1.4	[FCR 19] If process controls are used to establish bounds on flaw sizes in fracture critical parts, the approach shall be subject to the following: a. The approach is documented in the FCP. b. The rationale establishing that the approach is applicable is documented in the FCP. c. The FCP is approved by the RFCB.		
	8.1.5	[FCR 20] Spaceflight hardware with detected flaws that is used for flight without being repaired or replaced shall have a specific detailed assessment approach documented with rationale in the FCP that contains the following: a. An assessment approach of metallic parts by analysis is to include the following items in addition to the items in section 7.3.2 in this NASA Technical Standard: (1) Upper bound flaw size. (2) Upper bound crack growth rate. (3) Lower bound critical stress intensity factor or residual strength.		

		NASA-STD-5019A		
Description	Section	Requirement in this Standard	Applicable (Yes or No)	If No, Enter Rationale
		(4) Lower bound cyclic fatigue crack growth threshold stress intensity range (ΔK_{th}). b. An assessment approach for composite or bonded parts with detected flaws is to include the following items: (1) The approach and rationale provided to the RFCB for approval before implementation. (2) Documentation of the approved approach in the FCP.		
	8.2	[FCR 21] Traceability for each fracture critical and NFC composite or bonded part shall be established and maintained by providing a unique serial number (or other method when serialization is not practical) and a complete life history, including load history, impact damage, repair, materials, manufacturing, processing, and environmental exposure.		
	8.3	[FCR 22] The selection, processing, and use of materials for all fracture critical and NFC composite or bonded parts shall include the following items, which are documented directly in the FCSR or the items have pertinent documents referenced in the FCSR: a. Fabricate parts from materials with supplier data certifications. b. Select materials compatible with NASA-approved Standards and Specifications. c. Account for the effect of operating conditions on damage tolerance properties. d. Design and assess with strength and damage tolerance properties that		

		NASA-STD-5019A		
Description	**Section**	**Requirement in this Standard**	**Applicable (Yes or No)**	**If No, Enter Rationale**
		are generated by tests on samples representative of the flight hardware material, subject to either item (1) or (2) below: (1) Material is processed to the same thickness, material process condition, and material orientation in the part) that result in the worst combination for damage tolerant assessment. (2) The material process condition and the material orientation are fully traceable throughout fabrication and service life. e. Derived strength and damage tolerance data obtained from NASA-approved sources. If data are lacking, data are conservatively bounded or determined by sufficient testing to assess scatter to provide averages with testing approved by the RFCB. f. Obtain an approved MUA for any materials not developed and qualified in accordance with the requirements of NASA-STD-6016. g. Include all MUAs in the FCSR.		

Table 7—Documentation and Verification

NASA-STD-5019A				
Description	**Section**	**Requirement in this Standard**	**Applicable (Yes or No)**	**If No, Enter Rationale**
Documentation	9.1.2	[FCR 23] Fracture critical parts shall be identified on engineering drawings in the notes of the individual part drawing, along with the inspections, serialization, and other pertinent information necessary to maintain traceability of the part and its history of manufacturing and use.		
	9.1.3	[FCR 24] An FCSR shall be developed by the spaceflight hardware program or project that: a. Documents the basis for acceptance that all the flight hardware parts have met the fracture control requirements in the approved FCP. b. Contains detailed information or reference to detailed information for all parts, including results for evaluations, classification, assessments, inspections and other pertinent records, and their disposition for fracture. c. Documents all assessments, such as analyses and tests, conducted on representative flight hardware used for flight certification. d. Identifies the flaws and impact damage threats that are accepted on risk by the program authority, i.e., the flaws and impact damage threats for which there is no damage tolerance evaluation. e. Is approved by the RFCB.		
Verification	9.2	[FCR 25] Verification of adherence of the flight hardware to the fracture control requirements in this NASA Technical Standard shall include all of the following:		

NASA-STD-5019A				
Description	**Section**	**Requirement in this Standard**	**Applicable (Yes or No)**	**If No, Enter Rationale**
		a. Written documentation that establishes that each requirement has been met. This documentation describes how the requirement was verified, e.g., test, analysis, inspection. The project is responsible for providing this verification, including assurance that fracture control activities were implemented on the flight hardware before flight and re-flight, to the appropriate program management. b. Approval of the FCP and FCSR by the RFCB, documented by a concurrence memorandum from the RFCB to the applicable project/program office. c. In the event of conflict between the RFCB and the applicable project office concerning verification of compliance with fracture control requirements, follow the procedures in place at each NASA Center to resolve technical conflict, with the option to appeal to the NASA Chief Engineer for final resolution.		

Table 8—Alternative Approach Requirement

NASA-STD-5019A				
Description	Section	Requirement in this Standard	Applicable (Yes or No)	If No, Enter Rationale
Alternative Approaches	10	[FCR 26] If alternative approaches are proposed (rather than meeting any part of the accepted approaches that are prescribed in sections 5, 6, 7, or 8 in this NASA Technical Standard, with the exclusions show below), the alternative approach shall include all of the following items: a. Provide an equivalent assurance of mitigating the risk of catastrophic failure from flaws during the service life of the hardware. b. Have the approval of the RFCB. c. Meet all the other applicable requirements in this NASA Technical Standard. d. FCRs 10, 15, 20, 21, and 22 (sections 7.1, 8, 8.15, 8.2, and 8.3, respectively, in this NASA Technical Standard) are excluded from alternative approach consideration. *Note that FCR 26 pertains to sections 5, 6, 7, and 8 only; therefore, FCRs 1, 2, 3, 4, 5, 23, 24, and 25 are also excluded from alternative approach consideration.*		

APPENDIX B

REFERENCE DOCUMENTS

B.1 Purpose and/or Scope

The purpose of this appendix is to identify relevant guidance documents for application of this NASA Technical Standard. This is not intended to be an exhaustive listing of all the documents the hardware developer may find useful in implementing this NASA Technical Standard. The latest issuances of cited documents apply unless specific versions are designated.

B.2 Government Documents

Department of Defense

MIL-HDBK-6870	Nondestructive Inspection Program Requirements for Aircraft and Missile Materials and Parts

DOT

DOT Title 49	Code of Federal Regulations Title 49, Transportation
MMPDS	Metallic Material Properties Development and Standardization

NASA[8]

NASA-HDBK-5010	Fracture Control Implementation Handbook for Payloads, Experiments, and Similar Hardware
MSFC-RQMT-3479	Fracture Control Requirements for Composite and Bonded Vehicle and Payload Structures
NPR 7120.5	NASA Space Flight Program and Project Management Handbook
PRC-6509	Process Specification for Eddy Current Inspection

---

[8] NASA-HDBK-5010A is under development and may not be released at the time of publication of this NASA Technical Standard. Before the release of NASA-HDBK-5010A, the current handbook can provide interim guidance for applying this NASA Technical Standard.

B.3 Non-Government Documents

AIA/ NAS

NA0026	Procurement Specification Metric Fasteners, A-286 CRES Externally Threaded, 1100 MPa Tensile, 660 MPa Shear
NA0271	Metric Fasteners, CRES 300 Series, Externally Threaded, MJ Thread, 500 MPa F_{tu} and 700 MPa F_{tu}
NAS4003	Fastener, A286 Corrosion Resistant Alloy, Externally Threaded, 160 KSI F_{tu}, 95 KSI F_{su}, 1000 °F
NASM85604	Bolt, Nickel Alloy 718, Tension, High Strength, 125 KSI F_{su} and 220 KSI F_{tu}, High Temperature, Spline Drive

American Petroleum Institute (API)/American Society of Mechanical Engineers (ASME)

API 579-1/ASME FFS-1	Fitness-For-Service

ASME

ASME BPVC-VIII-1	Boiler Pressure Vessel Code, Section VIII, Rules for Construction of Pressure Vessels Division 1
ASME BPVC-VIII-2	Boiler Pressure Vessel Code, Section VIII, Rules for Construction of Pressure Vessels Division 2-Alternative Rules

ASTM International (formerly American Society for Testing and Materials)

ASTM E399	Standard Test Method for Linear-Elastic Plane-Strain Fracture Toughness K_{Ic} of Metallic Materials
ASTM E561	Standard Test Method for K-R Curve Determination
ASTM E1681	Standard Test Method for Determining Threshold Stress Intensity Factor for Environment-Assisted Cracking of Metallic Materials
ASTM E740/E740M	Standard Practice for Fracture Testing with Surface-Crack Tension Specimens
ASTM E1820	Standard Test Method for Measurement of Fracture Toughness

| ASTM E1823 | Standard Terminology Relating to Fatigue and Fracture Testing |
| ASTM E2899 | Standard Test Method for Measurement of Initiation Toughness in Surface Cracks Under Tension and Bending |

SAE

| AS7468 | Bolts, Cobalt-Chromium-Nickel Alloy, UNS R30035, Tensile Strength 260 Ksi, Procurement Specification |

B.4   Other Documents

Boyer, R.R.; Spurr, W.F. (January 1978). "Characteristics of Sustained-Load Cracking and Hydrogen Effects in Ti-6Al-4V," *Metallurgical Transactions A.* Vol. 9A, pp. 23-29.

Lewis, J.C.; Kenny, J.T. (July 1976). *Sustained Load Crack Growth Design Data for Ti-6Al-4V Titanium Alloy Tanks Containing Hydrazine.* Paper presented at AIAA/SAE 12th Propulsion Conference. Palo Alto, CA.